P9-DTN-027

Copper: Its Mining and Use
by the
Aborigines
of the
Lake Superior Region

Report of the McDonald-Massee Isle Royale Expedition 1928

By

Geo. A. West

Introduction to the Greenwood reprint by
ROBERT E. RITZENTHALER
Curator of Anthropology
Milwaukee Public Museum

GREENWOOD PRESS, PUBLISHERS
WESTPORT, CONNECTICUT

Originally published in 1929
by order of the Board of Trustees,
Milwaukee Public Museum, Milwaukee, Wisconsin

Reprinted with the permission
of the Milwaukee Public Museum

First Greenwood Reprinting 1970

Library of Congress Catalogue Card Number 76-111402

SBN 8371-4634-8

Printed in the United States of America

Introduction to the Reprint Edition

An important contribution of George A. West, a Milwaukee attorney whose serious avocation was archaeology, was the delineating and documenting of copper mining techniques, particularly as practiced by the prehistoric Indians on Isle Royale, Michigan. The importance of these early miners and fabricators of copper of the Upper Great Lakes Region is recognized in the name given to their time and culture, the Old Copper Indians.

An expedition to Isle Royale in 1928 provided data which made it possible to reconstruct the circumstances under which the Indians extracted the native copper they used in making their extensive tool kit of copper implements and weapons. Of equal importance, perhaps, is the discussing, typing, and illustrating of the implements as analyzed from the very extensive Old Copper collection in the Milwaukee Public Museum. While West included a few items from other copper-using cultures, such as the Hopwellian, about ninety-eight percent of his work was concerned with the Old Copper people.

A considerable amount of new information on this culture has come forth since the publication of this volume in 1929. At the time of West's publication all that was known was based on copper artifacts found on the surface. In the Wisconsin-Michigan area, four old Copper cemeteries have since been discovered and excavated by professional archeologists. We now have information on their burial practices, their lithic assemblage, and their objects made of bone, antler, and shell. We also have a series of radiocarbon dates placing them within a period of about 3,000 to 500 B.C. Two important sites near Ottawa, Canada, have recently been excavated and have provided further data.

Mr. West was a member of the Board of Trustees of the Milwaukee Public Museum from 1906-1938, and served as its president for seventeen years. His work, originally published by the Museum, deserves space in the library of anyone interested in this imaginative group of people, the earliest metal workers in the New World.

—Robert E. Ritzenthaler
Curator of Anthropology
Milwaukee Public Museum
Milwaukee, Wisconsin, 1970

Further References

DRIER and DU TEMPLE
1961 *Prehistoric Copper Mining in the Lake Superior Region.* (Published privately.)

GRIFFIN, JAMES B.
1961 "Lake Superior Copper and the Indians," *Anthropological Papers,* No. 17, Museum of Anthropology, University of Michigan, Ann Arbor.

HRUSKA, ROBERT
1967 "The Riverside Site," *Wisconsin Archeologist,* N.S., Vol. 48, No. 3 (September).

QUIMBY, GEORGE I.
1960 *Indian Life in the Upper Great Lakes, 11,000 B.C. to A.D. 1800.* Chicago, University of Chicago Press. (See pp. 52-63.)

RITZENTHALER, ROBERT E. (Ed.)
1957 "The Old Copper Culture of Wisconsin," *Wisconsin Archeologist,* N.S., Vol. 38, No. 4 (December).

WITTRY, WARREN
1951 "A Preliminary Study of the Old Copper Complex," *Wisconsin Archeologist,* N.S., Vol. 32, No. 1 (March).

WITTRY, WARREN and ROBERT E. RITZENTHALER
1956 "The Old Copper Complex, an Archaic Manifestation in Wisconsin," *American Antiquity,* Vol. 21, No. 3.

BULLETIN
OF THE
PUBLIC MUSEUM OF THE CITY OF MILWAUKEE

Vol. 10, No. 1, Pp. 1-182, Plates 1-30, Figs. 1-12, Maps 1-2 May 29, 1929

Copper: Its Mining and Use by the Aborigines of the Lake Superior Region

---

Report of the McDonald-Massee Isle Royale Expedition 1928

By
Geo. A. West

CONTENTS

PART I—THE McDONALD-MASSEE ISLE ROYALE EXPEDITION, 1928

PART II—PREHISTORIC COPPER MINING

PART III—ABORIGINAL COPPER ARTIFACTS

Page

ACKNOWLEDGEMENTS

To Commander E. F. McDonald, Jr., President of the Zenith Radio Corporation, and to Mr. Burt A. Massee, Vice-President of the Colgate-Palmolive-Peet Company, both of Chicago, are due the sincere thanks of the Milwaukee Public Museum and the writer for their generosity in making the McDonald-Massee 1928 Isle Royale Expedition possible, as well as for their valuable assistance in the field.

A special credit is also due to Mr. George R. Fox, Archeologist and Director of The Edward K. Warren Foundation, of Three Oaks, Michigan; to Professor Baker Brownell, Professor of Contemporary Thought at Northwestern University, Evanston, Illinois; and to Dr. Alvin LaForge, of Chicago, for their hearty cooperation and distinguished services in our archeological research work on the island, as well as for information furnished the writer.

In the preparation of this paper, the rich collections of the Milwaukee Public Museum, as well as those of other museums and private collectors were carefully studied. The writer takes this occasion to express his grateful acknowledgment and thanks for the aid afforded him, and among those whom he desires to particularly mention are:

Mr. W. de C. Ravenel, Administrative Assistant to the Secretary of the Smithsonian Institution, Washington, D. C.

Mr. S. C. Simms, Director, Field Museum of Natural History, Chicago, Ill.

Mr. P. O. Fryklund, Roseau, Minnesota.

Dr. J. Alden Mason, Curator of the American Section, Museum of the University of Pennsylvania, Philadelphia, Pa.

Dr. W. B. Hinsdale, University of Michigan, Ann Arbor, Michigan.

Mr. Frank M. Warren, Minneapolis, Minnesota.

Dr. U. C. Nelson, Curator of Archeology, American Museum of Natural History, New York.

Mr. Willoughby M. Babcock, Curator, Minnesota Historical Socity, St. Paul, Minnesota.

Dr. Geo. L. Collie, Logan Museum, Beloit, Wis.

Mr. Vetal Winn, Milwaukee, Wis.

Mr. Phillip Schupp, Jr., Chicago, Illinois.

Mr. Lee R. Whitney, Milwaukee, Wis.
Mr. E. F. Richter, Milwaukee, Wis.
Mr. C. L. Harrington, Supt. of Forests and Parks, Wis. Conservation Commission, Madison, Wis.
Mr. Raphael Zon, U. S. Department of Agriculture, University Farm, St. Paul, Minnesota.
Mr. Francis Dayton, New London, Wis.
Mr. Joseph Ringeisen, Jr., Milwaukee, Wis.
Mr. Towne L. Miller, Honorary Curator of Archeology, Milwaukee Public Museum, Fairwater, Wis.
Mr. J. P. Schumacher, Green Bay, Wis.
Mr. Emmit Scott, La Porte, Indiana.
Mr. J. T. Reeder, Houghton, Michigan.
Mr. Arthur C. Neville, Director, Neville Public Museum, Green Bay, Wisconsin.
Mr. E. A. Seglem, Chicago, Ill.
Oshkosh Public Museum, Oshkosh, Wis.

The writer's sincere thanks are especially due to Mr. Charles E. Brown of the Wisconsin State Historical Society Museum and Secretary of the Wisconsin Archeological Society, and to Messrs. Ira Edwards, Curator of Geology, W. C. McKern, Associate Curator of Anthropology, and Huron H. Smith, Curator of Botany, all of the Milwaukee Public Museum, for special assistance.

It should not be supposed, however, that these kind friends are in any way responsible for expressions or conclusions in the body of this Bulletin, unless so quoted.

ILLUSTRATIONS

Plates

Maps

Text Figures

Part I

THE McDONALD-MASSEE ISLE ROYALE EXPEDITION 1928

THE TOPOGRAPHY OF THE ISLAND

The very extensive aboriginal copper workings of the Lake Superior District are divided into two great groups,—one on Isle Royale and adjacent islands; the other, less remote, located on the south shore of the lake.

Isle Royale (Minong), located in the northwestern part of Lake Superior, is forty-four miles long with an average width of about four and a half miles, most of its shore line being protected by a chain of more than a hundred and fifty small islets and many minor rocks.

The topography of the island is rough and rugged. It is traversed by two principal ranges, one known as the Greenstone Range, the other the Minong Trap Range. The Greenstone Range extends the entire length of the island, rises to a height of four hundred and fifty feet and forms the divide. The Minong Trap Range, which lies about a mile distant from and parallel to the Greenstone, with an elevation of approximately four hundred feet, is second in importance. Between these ranges lie five fairly large lakes and many small ones.

At a very early time, successive lava flows covered the island. These were later broken up by dykes and faults, and acted upon by glaciers and erosion, leaving it a land of high ridges, deep valleys and rocky headlands. The great glaciers, coming from the north, cut down the copper-bearing rock of the island hundreds if not thousands of feet, filling in its low places with this and other glacial material. The southward flow of the ice stream continued for many centuries and then slowly melted back. The retreating ice dammed up the waters of Lake Superior to a great depth, as old beach lines on the island, found more than four hundred feet above the present level of the lake, clearly indicate.

The rocky coast with its fjord-like channels and numerous reefs makes access to the island by boat not only difficult, but exceedingly dangerous to the stranger. Where the depth of soil permits, there usually exists a perfect jungle of stunted trees, shrubs and plant life. The rocks are generally moss-covered, making exploration no easy task. There still remain a few scattered strips of pine and hardwood of some commercial value that escaped the devastating fires of the last century.

THE GEOLOGY OF ISLE ROYALE

Because of its location, Isle Royale today presents an example, almost unequaled, of nature's handiwork in building up and tearing down rock structures. A visit to this ancient island is bound to create an interest in the observant, and awaken a desire for information as to its origin and the vicissitudes through which it has passed.

The writer, therefore, presents a concise and comprehensive statement of the geological history of the district in which the island is located, furnished by Mr. Ira Edwards, Curator of Geology, Milwaukee Public Museum, who was a member of its Isle Royale Expedition, 1924.

"The rocks of Isle Royale are exceedingly ancient, having been formed in that portion of geologic time known as the pre-Cambrian era which includes that exceedingly long interval before the existence of life on the surface of the earth. Even if life had been present we could not expect to find any remains as fossils, in the rocks of this island, as these are almost entirely lavas of volcanic origin. They are interbedded with smaller amounts of sandstone and other fragmental materials. These lavas are of that group best known as trap rocks but more exactly as gabbro, melaphyr and dolerite, depending upon the coarseness of the crystals. They did not issue from a volcano in the ordinary sense of the word for very little volcanic ash or other evidence of explosive action is present. They seem to have come from an open fissure or low dome rather than from a high cone such as that now present at the volcano, Vesuvius. At any rate the lavas spread far and wide over the western part of the Lake Superior Basin reaching westward into Minnesota and southward into Wisconsin and Michigan. Flow after flow spread over the same region each adding its bulk to that of those which had preceded it. As a result of this action we have today a great thickness of lava covering wide areas in the vicinity of Lake Superior.

"Between eruptions of lava there seem to have been periods of quiet when layers of sandstone and conglomerate accumulated on top of the lava flows. In the early part of the period the lava outflows occurred at short intervals and the sandstone beds between them are thin and relatively insignificant. But toward the close of the epoch the periods of quiet became more numerous and the sandstone accumulated in greater and greater quantities, until at last the lava ceased to flow and the accumulation of sediments went on uninterrupted. On Isle Royale the sandstone beds are not very important but on the southern shore of Lake Superior they form a great portion of the mass of rock which can be assigned to this section of the geologic column.

"When the lava flows came to the surface, they contained great quantities of gaseous matter which was actively given off from the surface of the liquid lava. As a result, the upper portions of these flows became more or less porous and in many cases contained myriads of large or small holes, usually oval in shape, left by the escape of the contained gases. Lower down in the same flow, the gas was unable to make its way to the surface with such ease and was given off much slower so that the openings become fewer and smaller as one descends into the mass. At the base, there is usually found lava which is relatively dense and without openings of any noticeable size. This gradation in character from top to bottom of the various lava flows which make up this series of rocks, is very constant. The upper portion has received the special name of amygdaloid to distinguish it from the more ordinary-appearing rock below. There is not only this gradation in porosity but also a similar gradation in coarseness of grain, the finer-grained portions occurring near the top and the coarser-grained rocks near the base where the cooling was relatively slow. At the very base of the flows, there is sometimes found a conglomerate made up of fragments of the surface rock which were picked up by the advancing lava and engulfed in its mass.

"All of the various rock varieties before mentioned can be found on Isle Royale, but most of the rock outcrops are made up of the lower, denser portions of the lava flows which are much more resistant to weathering. These now stand up above the general surface as hills and ridges. The upper amygdaloidal portions are more often exposed along the cliffs on the shore line of the island or on the shores of the numerous lakes in its interior. The cliffs bordering on the fjord-like coves on the north side of the island, also afford numerous exposures of the lava. The sandstones interbedded between the lavas are seldom seen at the surface in this region.

"Some time after the cessation of the lava flows but before the end of the deposition of the sandstone, the rocks of the Lake Superior District were involved in a movement of the earth's crust which completely altered their position. Those in the western part of the Lake Superior Basin were bent into a great, downfolded arch or syncline which is practically outlined by the present shore of the lake. The north edge of this syncline is seen in the rock cliffs of the Minnesota Coast and Isle Royale. The south edge forms the ridges of the Keweenaw Peninsula and the high land of northern Wisconsin. Thus it is that the same rocks which occur in Isle Royale, after disappearing beneath the surface of Lake Superior, come to light again on the Keweenaw Peninsula. It is also the reason why the cliffs surrounding Lake Superior all face away from the lake, those on the north side facing northward and those on the south shore facing southward. This arch also accounts for the fact that the rocks of this region always dip toward the lake. This dip varies greatly in different places from nearly vertical to a gentle slope. The rocks on Isle Royale dip at a low angle to the southward.

"We do not know exactly how many lava flows were involved in this movement nor do we know the number which are represented by the rocks on Isle Royale. At least seven different outflows of lava occur on the island. These are indicated by the seven principal ridges which make up the surface of the island and the neighboring groups of smaller islands. As would be expected, these are parallel to the greatest length of the island and each individual ridge extends with more or less interruption throughout the entire length of the island. The numerous lakes in the interior lie in the valleys between these prominent ridges and the short streams which drain them break through the ridges in more or less steep-walled gorges.

"During or after the folding of this region, the rocks were broken by a series of faults which have caused a greater or less interruption in the continuity of the various lava flows. Following this period of movement, there was an exceedingly long time, covering almost the entire period of geologic history, during which these pre-Cambrian rocks were exposed to the action of the weather. During this time the surface features which we now recognize in this region were produced. It is probable that this period of erosion was interrupted at various times by temporary submergence in invading sea waters. If so, there is today no record of the extent of these invasions of the sea.

"It must not be thought that these rocks are uninteresting because of the absence of fossil remains. The minerals which they contain more than make up for the lack of fossils. These are found mostly in the amygdaloidal upper portions of the lava flows. The cavities left in the rock by the escape of gases have been completely filled by the deposition of minerals within them. The manner in which this deposition was brought about is not very clear, but it must have been by means of waters carrying mineral matter in solution, which seeped through the porous rock and left the mineral matter behind. Probably these were hot waters coming up from greater depth, but some deposition might have been caused by surface waters seeping downward.

"Of these various minerals the most important though least spectacular, is copper. The great deposits of this mineral which occur on the Keweenaw Peninsula and in Ontonagon County, Michigan, are too well-known to need further comment. They have supplied the material for a great mineral industry which has flourished for a period of nearly three-quarters of a century. The lavas on Isle Royale also carry the copper but not in a quantity which has proven profitable for mining. The various sporadic attempts to mine Isle Royale copper have all ended in failure and there has been no active mining in this district for many years. Intimately mixed with the copper occurs native silver, a combination almost unknown in any other mining district. This curiously colored ore has been given the name of 'half-breed' by the miners.

"Along with the ore many other minerals were deposited in the cavities of the amygdaloidal rock. The commonest of these are calcite and the various minerals of the zeolite group; analcite, datolite, stilbite, chabazite and others. This group of minerals is quite common in situations of this kind occurring in similar lava flows in New Jersey and Colorado, as well as many other places throughout the world. The very beautiful crystal groups which are to be found make a good appearance in any mineral collection and the search for them is interesting to the collector. Beside these more common minerals, the lavas of Isle Royale have yielded two varieties which are unknown outside of the Lake Superior District, thomsonite and chlorastrolite. Thomsonite is found in rather small masses in the openings in the amygdaloids and is of a beautiful light green or light pink tint. It is usually banded, the markings being roughly parallel to the sides of the cavity which the individual specimen fills. Chlorastrolite is a very dark green mineral often so marked with black lines that it appears to be broken up into fragments cemented together by a black material. It is not known in as large nodules as thomsonite and is usually more regular in outline. Neither of these minerals has ever been found in a crystalline condition. Both are very resistant to the weather and have accumulated on the beaches surrounding Isle Royale where they are much sought after by mineral collectors and others. Both of them have been used to a limited extent in jewelry and rank among the more unique if not among the most precious stones.

"The long period of weathering and erosion was brought to an end by the invasion of the first of the continental ice sheets which characterized Pleistocene time in North America. It is not known just exactly how many of the ice sheets covered the Lake Superior Region, but it is probable that, at least, the last two extended over the entire lake basin including, of course, Isle Royale. This northern country was relatively near the region of ice accumulation which was along the shore of Hudson Bay and consequently was very heavily glaciated by an exceedingly thick load of ice. This material not only removed all of the products of erosion which were in its path but tore down the hard rock surfaces as well. It rounded all of the hills and ridges on Isle Royale and left a record of its passage in striations on the surface of the rock. The island lay directly across the path of progress of the ice and its valleys were greatly deepened.

"After the ice sheet had reached its maximum a recession set in which eventually cleared the country of the glacier. During its retreat the melting ice sheet left behind all of the material which the ice had picked up in its forward advance. It happens however, that the main masses of this glacial debris were transported much further south than the region of Isle Royale, and the various accumulations of glacial material which are present on the island are relatively thin and do not conceal the previous contour of the bed rock surface. This is also probably due in part to the fact that the recession was rapid during the period when the ice front was in the vicinity of Isle Royale. There is left over the island only scattered piles of glacial boulders and a small amount of ordinary glacial till. This material was brought from some point in Canada where the rock formations were of an entirely different formation from the lavas of the island. These boulders are often of granite or some other coarse-grained igneous rock and served as excellent hammerstones for the aboriginal miners who came to the island in search of copper. The hundreds of hammerstones now found in the numerous pits of these ancient workers are almost without exception glacial rocks, of materials foreign to the island.

"The present Great Lakes drainage system came into being as soon as the ice cover melted but at first the outlet by way of the St. Lawrence River was blocked by the ice, and the waters were forced to find a higher outlet. It is not necessary here to go into a discussion of the various outlets and levels of the glacial lakes but it is sufficient to note that the waters of Lake Algonquin, the highest in the Lake Superior Basin, entirely submerged Isle Royale and the neighboring islands. Little by little the ice melted and opened lower and lower channels by way of the St. Lawrence Valley. The water in the lake gradually subsided to accommodate itself to these new outlets. Each recession of the ice was thus accompanied by a lowering of the level of the lake and the formation of a new beach line. Many of these old beach levels left in this way high up on the slope of the hills of Isle Royale can still be seen but there has been no systematic mapping of any of them in this region. It can be said however, that nearly all of the gravel deposits to be found on the island are connected with a beach level of a former lake. Some typical rock cuttings occur on the beaches but, in general, the rocks of Isle Royale are so resistant to wave action that they were little affected by the waves during the short period at which the lake remained at any given level. The construction of gravel bars built largely of the material of the glacial deposits was not so difficult and was accomplished in a much shorter time. They are consequently the chief indications of the existence of the former lake levels and may most easily be recognized by the fact that the upper surfaces of the deposits remain at a given level over a considerable extent of territory. It is true that there has been a tilting of the land surface and a consequent tilting of these old lake levels since their abandonment but this is not of a sufficient degree to interfere with the apparent horizontality of the beach lines. It can only be ascertained by following the beaches for a long distance.

"This tilting is responsible for the last of the steps in the geologic history of this country. When the ice sheet left the St. Lawrence Valley the land lay at such a low level that the ocean waters entered the Lake Ontario Basin and the upper lakes were drained by way of Lake Nipissing and the Ottawa River. Subsequent tilting of the land was greatest in the northeast with the result that the lower portion of the Great Lakes Region was brought up above the surface of the sea and the outlet by way of Lake Nipissing was no longer practicable. The lake waters then found their present outlet by way of the Detroit River. This movement continued for a long time and in fact may still be in progress. If this is true, the St. Lawrence River is able to erode its channel as fast or faster than the uplift of its bed with the result that the water level of the lakes has been kept without change. It is noticeable that streams entering Lake Superior on the south and west shores have their lower portions flooded and their valleys occupied by marshes. This is not true of those streams which enter the lake on the north and east shores showing that the waters of the lake are gradually responding to the tilting of the land surface. This influence is not sufficient to have been felt in the small streams of Isle Royale but it may be responsible for the small marsh areas at the ends of the coves on the north shore of the island.

"At present geologic activity is not very noticeable. There are no streams of any considerable size which might perform notable erosion of their valleys and the only seat of geologic action is along the shore of Lake Superior. Here gravel beaches are being built and the cliffs are being destroyed by wave action. In recent times man has done much more to alter the appearance of the island than has been accomplished by geological forces."

HISTORICAL

Isle Royale was ceded by the Chippewa Indians in 1843, after which it was invaded by prospectors and explorers so that in 1847 it presented perhaps as lively a scene as ever in its history, but this period of activity was of short duration. In 1855, "the island was a desert once more with no permanent inhabitants". This passive condition of affairs lasted until 1867 when explorations were somewhat revived and quite extensive but unprofitable mining operations were carried on for several years in McCargoe Cove, Washington Harbor, and other parts of the island. A relapse followed in 1891, and all activity ceased in the following year.

During the latter part of the 18th and the early part of the 19th centuries, a number of fur trading posts were established on and about Isle Royale. They were, however, not interested in copper mining, as the securing of furs was far more profitable.

At one period of its history the island was organized as "Isle Royale County", with its county seat, called Ransom, located at the foot of Rock Harbor. The main attraction was its mineral resources, as the forest growth is too stunted and inaccessible to have merited the attention of lumbermen.

CLIMATIC CONDITIONS

We found the temperature of the waters of Lake Superior about the Isle Royale archipelago, 40 degrees Fahrenheit, readings being taken three feet below the surface.

The days of our sojourn were balmy, although along the coast, cool and bracing winds at times made warm clothing desirable. The thermometer seemed to fall with the setting sun to a point conducive of invigorating sleep.

During the summer season, the few resorts, located on its shores, are fairly well patronized; while in the winter, because of the severe cold, the island is uninhabited excepting by one or two game wardens who remain there to trap the coyote and lynx.

PERSONNEL OF THE EXPEDITION

On July 21, 1928, two palatial yachts, the "Naroca" and the "Margo", with booming cannons, gleaming with highly polished brass, bright red cushions on the quarter deck and shining mahogany, docked at the Milwaukee Yacht Club on their way to Isle Royale, in order to take aboard two gentlemen together with the writer, President of the Board of Trustees of the Milwaukee Public

Museum, the expedition being under its auspices. The "Naroca," (Plate IV, fig. 1), is owned by Commander E. F. McDonald, Jr. of Chicago, President of the Zenith Radio Corporation, who was with McMillan on two Arctic expeditions and who commanded the "Peary" on the last one. The "Margo" (Plate IV, fig. 2), belongs to Mr. Burt A. Massee, also of Chicago, Vice-President of the Colgate-Palmolive-Peet Company. These two gentlemen financed the expedition.

Aside from Commander McDonald and Mr. Massee, the accompanying members were Mr. Geo. R. Fox, Archeologist and Director of the Edward K. Warren Foundation, Three Oaks, Michigan; Professor Baker Brownell, Professor of Contemporary Thought, Northwestern University; Dr. Alvin LaForge, a prominent Chicago diagnostician, Mr. John Kellogg of Chicago, champion fly-caster of the State of Illinois; Mr. Joseph T. Reilly and Mr. John F. Dunphy, both of Milwaukee, the latter three being life-long friends of Mr. Massee; and the writer, whose specialty is archeology.

Mr. U. J. Hermann, owner of the schooner, "Swastika", sailed later to join the expedition, but encountered a water spout, which caused him to put into Manitowoc for repairs. As a result he joined us only a few hours before the "Margo" and the "Naroca" hoisted anchors, passed out of McCargoe Cove, down Amygdaloid Channel, and across the reef that guards its entrance, homeward bound.

OBJECTS OF THE EXPEDITION

The objects of the expedition were to determine, if possible, what peoples did the primitive copper mining on this mysterious island and when they worked; if any traces of occupancy by the Norsemen could be discovered; if a fortified city of "pit-dwellers", as had been reported, existed there; and also to locate and examine as many of its mines, caves and village sites, as time would permit, to the end that the culture of its prehistoric workers be determined. The plan of the expedition was to make a preliminary archeological survey, laying the groundwork for a future program of research and exploration, and to carry on the work so well commenced by the Milwaukee Public Museum at McCargoe Cove in 1924.

ON THE WAY

A stop was made at Sturgeon Bay where a supply of ice was taken aboard. After calling at Harbor Springs, we next arrived at beautiful Mackinac (Plate V, fig. 1). A tour of the island was taken and the following day was spent among the famous Snow Islands. There are about fifty of these, separated by narrow channels, and highly improved by charming summer residences, principally owned by wealthy citizens of southern Michigan and Indiana.

Our trip through Lake Huron was extremely rough yet we reached the historic Soo without difficulty (Plate V, fig. 2). A call was made on Mr. J. W. Turran, Editor of "The Evening News" at Sault Ste. Marie, Ontario, where we had the pleasure of viewing his archeological collection containing many interesting objects.

Leaving the Soo, high winds were encountered on Lake Superior and the "Margo", on which the writer was a guest, sought shelter in Harmony Bay on the north shore, where we remained throughout the day. However, time was not lost as an Indian village site and burying-ground, as well as the remains of an old trading post were located on the west bank of this charming body of water. As we passed out of the bay, a large buck deer stood on the nearby shore taking a farewell look at us.

The "Naroca" went on to Michipicoten Isle, which place we reached the following afternoon, when after a short stay we proceded direct to Rock Harbor, Isle Royale, again joining the "Naroca".

AT THE ISLAND

Here is located Rock Harbor Lodge, a very attractive summer resort (Plate VI, fig. 1). The first thing done was to engage, as a local pilot, Mr. J. R. Anderson of Toban Harbor, who has a well-earned reputation of being a fearless navigator and life-saver, and to hire, as guide, Mr. John Linklater, a very intelligent Indian, whose summer address is McCargoe Cove. These men proved themselves to be most efficient.

We first visited Chlorastrolite Beach, where a number of greenstones were found. Several ledges and other beaches on the island contain this semi-precious stone.

We next crossed the channel to the old Saginaw Mine, famous

not for the extent of its shafts and tunnels, nor for its drifts containing ice the year round, but from a tragedy enacted there. The story is, that forty-two miners of various nationalities worked this mine in the early fifties. A ship that was to arrive late in the fall with provisions for the company never reached the island. Hunger and cold drove the miners to distraction. They quarreled among themselves with the result that all were killed excepting two who escaped and managed to survive throughout the winter. It is reported that these two men later visited the island and related the tale. The corpses, as far as possible, were gathered up and found a resting place on an island across the channel. Markers are to be seen on some of the graves with names and dates still legible.

A cave was visited on the west side of the channel, about a mile from Rock Harbor Lodge (Plate VI, fig. 2). This little cavern is located about sixty yards back from the water's edge and thirty feet above its level. It is well concealed among the rocks and shrubbery, the approach being through a passage-way about thirty feet wide with walls on either side for a short distance. Its entrance, nine feet high, is in the form of an arch and extends back in one direction twenty-two feet. This cave which faces east, has a branch running southward about fifteen feet, ending in a narrow opening. Its walls and roof are smoked black. In trenching the accumulations on its floor to a depth of about three feet, we found, all the way down, stones showing the action of fire. Strata of charcoal were also encountered, indicating that it had been occupied for a considerable period of time in the past, but not as a permanent residence. The walls contained no picture-writing and no artifacts or potsherds were found. We named this interesting place, "Susan Cave", after the name of a young lady who guided us to the spot.

It seems quite certain that the island at no time has been inhabited by the ancient miners during the winter season. The red metal was the only possible attraction to the island and mining there by primitive methods would be impossible in frigid weather.

Upon visiting the site of the old county seat, we found where the clearing had been, now partly overgrown by brush and trees. Piles of rock and what appears to have been a root cellar are the only evidences that remain of what was at one time a prosperous little town (Plate VII, fig. 1).

LAKE RICHIE

Another attraction, reached from the lower end of Rock Harbor, over a torturous trail of about three miles, is Lake Richie, where numerous moose can be seen at any time. Our party made two visits to this lake for the purpose of photographing these animals.

This is a shallow lake, containing much aquatic vegetation and is surrounded by a growth of fair-sized timber. As we approached the water's edge, two large bull moose, with great flaring antlers, were feeding a hundred yards or more before us. It is interesting to see them almost completely submerge like submarines, remaining under water from two to three minutes.

In another part of the lake stood a cow and a calf, in fact, moose were seen in almost any direction we saw fit to look. Mr. Fox and Prof. Brownell followed around one side of the lake, finding many opportunities for the use of the camera.

Dr. LaForge and the writer took a trail, near the water's edge, in the other direction and soon spied a cow moose coming to meet us. Sitting quietly down, she came within fifteen feet of our camera before turning away.

A little farther on, a bull and cow moose were seen standing near the trail and later went into the lake and walked down the shore. After being photographed, each left the water and passed within three feet of the writer, who was seated on the ground partly concealed behind a small white birch tree (Plate VII, fig. 2).

Upon relating this occurrence to Mr. Edmund H. Heller, a naturalist of note, who acted as President Roosevelt's guide during his African hunts, and is now Director of the Washington Park Zoo, Milwaukee, he made the following statement which is quoted as a warning to others: "You took great chances, as the male moose and elk, especially during the fall of the year, are the most dangerous animals with which we have to deal. If they are suddenly confronted with what they suppose to be an enemy, they immediately attack and strike a deadly blow with their front feet."

HAY BAY

Our yachts next cast anchor at Hay Bay, about thirty miles south of Rock Harbor. We visited Point Houghton, ten miles to the south, where a most interesting ossuary deposit is located which we named the "Massee Rock Shelter" (Plate VIII, fig. 1). Our party consisted of Massee, Fox, LaForge, Brownell and the writer as McDonald remained aboard his yacht nursing a broken rib.

From Hay Bay it was necessary for us to cross Siskowit Bay in two launches. On account of heavy seas, we were unable to round the point, so landed and walked about a mile across a rocky ridge to Fishermen's Home, (Plate VIII, fig. 2). This is occupied during the summer season by Viking fishermen, who guided us to the rock shelter, which is located in Section 32, Township 64 North, Range 36 West, about a hundred yards from the water's edge in a dense, dark forest and is underneath an overhanging ledge of red sandstone, about ten feet in height.

As we peered into the walled-up cave through a small opening our gaze met a grinning skull, in an upright position, above a heap of human bones, partly concealed by the dust of ages. The dignity of this vigil of the dead caused us to halt and meditate. It stared directly at us with a smile of welcome. We were convinced that if it could, a cry of joy would break the silence of that wilderness, that succor had come at last, that the great rock that had imprisoned its mortal remains for two hundred years or more would now be rolled away and the "happy hunting grounds" of its fathers be its goal.

One of the rocks, forming part of the wall that closed the entrance to this cave, which we had to move, was estimated to weigh fully a ton. We removed from this shelter, the principal bones of at least twelve human beings. One skull found was well preserved, the others being fragmentary (Plate IX). They were not of white men as we all agreed and Dr. LaForge pointed out characteristics of the shin bones encountered, indicating that they are the remains of moccasin wearers. Mr. Fox discovered among the bones a finely worked knife or spear-head of chert, three inches in length, an indication that the deposit belongs to the Stone Age. The bones, most of them yellow and decayed, had the appearance of great antiquity. The fair state of preservation of a few can be accounted for, in part, by the fact that the place of deposit was well protected from dampness. That they were the remains of Chippewa Indians or of some tribe that antedated them is quite evident. As a large number of the small bones were missing it is possible that the remains were gathered up and deposited, after first having been suspended in the trees for some time, as was the custom of the Chippewa and Sioux.

It is quite natural that rock shelters and caves should be used as depositories for the dead, because of the difficulty, on this island, of finding soil of sufficient depth for graves above the water level.

This ossuary was discovered by the children of Mr. E. T. Seglem in 1908.

In 1921 Mr. E. A. Seglèm of Chicago, while on a fishing trip to Isle Royale, so he informs me, took from the deposit a single skull in a good state of preservation. He is also of the opinion that other fishermen carried off a few skulls from the same deposit, but the young men who discovered the remains were quite certain that but one skull had been removed from the ossuary and that by their uncle, Mr. Seglem of Chicago.

We believe this to be the first discovery of prehistoric human remains found there. From information obtained, we feel certain that many other mortuary deposits will be located.

After several hours of strenuous work, our Indian guide called, "Lunch". In addition to our sandwiches and coffee, John had secured from the fishermen a seven-pound trout, which he had dressed and broiled over a fire, Indian fashion. Each of us enjoyed a finely cooked slab of fresh fish on plates of birch bark he had prepared.

In the evening, on our return to the yachts at Hay Bay, a bull moose was seen swimming near the shore. Two boats with outboard motors, manned by sailors, put out to head off the moose, that he might be forced to swim across the bay, thus giving us a better opportunity of photographing him. The moose attempted to reach shore a number of times, but was intercepted by one or the other of the boats. Finally with its hair on end, its eyes "blazing fire", its horns waving from side to side, it took after one of the boats and gave it a race for a block or more. After a narrow escape, the sailors admitted having had all the moose experience they desired.

Later in the evening, the very powerful search-light of the "Naroca" was played on a moose in the water which caused it to race up and down the shore, attempting frequently to land, but his own shadow, projected by the light, terrified him.

This was the banner day for the fishermen of the expedition, the larder of the ship being well supplied with speckled beauties.

THE OLD MINE

From Hay Bay we were guided over an anceint trail for about three miles to an old working (Plate X, fig. 1), in the southwest corner of Section 22, Township 64 North, Range 37 West. This was previously visited by Mr. Wm. P. F. Ferguson, who called it, "The Old Mine".[1]

Here we observed two methods of prehistoric mining; one, the following of copper indications into the solid rock by means of fire, water and mauls; the other, the pitting and trenching of glacial deposits and old beach lines for float copper. At this place we found a pit, ten or twelve feet in depth, located at the base of a rocky hill and apparently in glacial deposits. The debris from this pit, as well as an ancient beach line, had been trenched by Ferguson. We observed a few other smaller pits nearby in the amygdaloid rock. Recent prospecting was in evidence, but copper indications seemed poor at this place. About two miles farther west, white men had an unsuccessful mining experience.

The long trenches that Ferguson reports were considered by us as natural depressions, although there were several short ones, probably dug for the purpose of uncovering the native rock and others made in the loose debris of glacial deposits in pursuit of float copper. Hammer stones and mauls which were doubtless taken from the remains of a nearby ancient beach line were numerous. These workings being extremely limited and somewhat disappointing did not seem to warrant the spending of much time in a search for the camp site of the ancient miners who did this work.

We were guided to the "Old Mine" by a game warden named Bill Active, who assisted Mr. Ferguson in his excavations. In going over the trail we passed a large moose wallow, showing much use. This was on the edge of an extensive swamp, in which we were informed a number of upland caribou still live. The saplings in many places were trimmed for twelve or fifteen feet above the ground, which Mr. Linklater explained was caused by the moose that in the winter season stradle the tree, push it down, and eat its tender branches. When the weight is removed, the tree

[1]Ferguson, Wm. P. F., Michigan History Magazine, Vol. 8, No. 4, p. 450.

straightens up. He also informed us that moose would invariably appropriate a newly blazed trail to their own use.

THE OLD TOWN AND OLD FORT DIGGINGS

Our principal object in going to Hay Bay was to carefully investigate the "Old Town" of pit-dwellers and the "Old Fort Diggings", located up the Sibly River, about half a mile from the bay, in sections 23 and 24, Township 64 North, Range 37 West, as described by Mr. Wm. P. F. Ferguson[2] who refers to the two groups as "pit-dwellings", one of them surrounded by a clearly traceable earth work, etc.

The "Old Town", located on the southwest side of the river, consists of about seventeen oblong pits and several trenches, most of which were dug into ancient beach line deposits (Plate X, fig. 2).

The ridge on which the mining was done is composed of gravel and rounded boulders of considerable size, in which we found specimens of the red metal. We inspected pits excavated by Mr. Ferguson and did much digging ourselves. Dr. LaForge and Prof. Brownell completely cleared out one large pit, finding no walls or fireplaces, as Ferguson thought he did in others of this group.

Our conclusions were that the walls found by Ferguson in excavating pits were merely boulders laid up by the ancient diggers, in order to get them out of the way. The finding of fireplaces and charcoal in them, as Ferguson reports, might indicate that the ancient miners cooked and ate their meals in some of these excavations, or that they were temporarily occupied by the miners as a protection against cold winds, but not as permanent residences nor basements for the same. Successive forest fires in centuries past may have distributed charcoal throughout the various strata of soil accumulations found in the pits. Such fires are sometimes very hot and might crumble rock with which they come in contact, even in the bottom of a pit. Consequently, the finding of charcoal or stones, showing the effects of fire in mines of this class means but little.

These pits are naturally oblong instead of round, as they are merely trenches carried in and down until the disposal of the material became troublesome or the "dirt" ceased to pay. "That the embankments surrounding the pits were made to exclude surface

[2]Ferguson, Wm. P. F., Michigan History Magazine, Vol. 7, Nos. 3 and 4; Vol. 8, No. 4.

water", as Ferguson says, is out of the question as they are merely the natural heaps of earth and rock thrown out, and appear to a greater or less extent around all ancient excavations on the island. There was no necessity for moving back the debris. In the case of the "Old Town", the diggings are in the crest of a ridge and needed little or no protection against surface water. Mr. Ferguson says of these pits, "they are rectangular instead of circular like the mines". The result of the writer's observation is that the mines at McCargoe Cove, which are in solid rock as a rule, are sometimes round but usually oblong, frequently extending into the ledge for many feet,—all depending on the direction of the little streak of malachite followed.

Mauls so plentiful on most aboriginal mining sites were conspicuous in the "Old Town" by their almost total absence; there being no occasion for their use in working glacial or beach line deposits. Considerable work was done here and there is no evidence of white-man's intrusion. That the pits and drifts at this place were made solely in search of float copper and not for dwelling places seems certain.

THE OLD FORT

This interesting site is situated two hundred yards northeast from the "Old Town" and across the turbulant little Sibly. It consists of a rectangular tract, approximately a hundred and ten feet long and sixty feet wide, surrounded on one end and side by natural ledges of trap rock, ten feet high; on the other side and end by an ancient beach line of about the same height and fully thirty feet wide (Plate XI, fig. 1). Although Mr. Ferguson considered these embankments artificial or partly so and built for protection, calling the enclosure the "Old Fort", we found no evidence warranting such a conclusion. Within these walls, especially along the ledge of trap rock, glacial deposits exist to a depth of ten or more feet.

About two hundred yards farther north we found another enclosure, similar in formation, with perfect angles and of practically the same size. Within the "Old Fort" there are nine shallow pits and near the ledge of trap rock, one large one, thirty feet long, fifteen feet wide and fully ten feet deep. This was partly excavated by Ferguson, who found what he thought to be laid-up walls with a fireplace and other evidences, causing him to conclude that it was a "pit-dwelling" (Plate XI, fig. 2). On the surface within the enclosure was found a small dump of copper-bearing rock, probably on the spot where the copper was separated from chunks of rock. In the soil below this dump, Mr. Massee found a very beautifully formed chalcedony arrow-point.

Mr. Fox observed a small opening within the enclosure, underneath the lava flow, about forty feet west of the large pit and at once began excavating at this point. He went down five feet without reaching the original floor, and drifted at least six feet under the trap rock, encountering broken mauls and evidences of fire all the way. It was also found that the under-part of this lava flow or ledge was copper-bearing and partly chipped away. Further investigation proved that the drifting under the rock had continued for fully a hundred and fifty feet of its length, passing through the edge of the large pit, beyond it, and across the ancient beach line. Thus was discovered a unique prehistoric mine,—nothing of the kind having heretofore been found on the island (Plate XII, fig. 1). The discovery and further excavation of the large pit, with its "walled-up sides", led to the conclusion that it was not a "pit-dwelling", but dug for the purpose of getting under the copper-bearing ledge; and that because of its being in glacial deposits, it was enlarged and elongated in search of float copper.

The boulders laid up on its sides, some of which weighed hundreds of pounds, were evidently placed there to dispose of them in the easiest possible manner. In relation to the method in use by these early miners, J. H. Forster says, "they did not carry the rock out to the surface to dump it, but piled it up neatly on each side of the drift".[8] Had the material from the large pit been removed by the previous excavations for a few feet more on the side toward the ledge, the mining underneath it would have been discovered and the object of the construction of the pit made evident. We believe that the authors of these extensive works had their camping grounds across the stream a short distance from the "Old Town" at a place we noted as being well adapted to the purpose and which deserves careful investigation (Plate XII, fig. 2).

[8] Forster, J. H., Rep't. Smithsonian Institution, 1892, p. 182.

WRIGHT'S ISLAND

The yachts anchored at Wright's Island for the night. Here, Mr. Charles Purdy, one of the resident fishermen, fed the gulls offal from the fishing operations, in order that an opportunity be given for photographing them. Although less than a dozen were in sight when the food was first dumped, hundreds of them appeared in a few moments. While there, two lake trout weighing nearly forty pounds each, which had been caught on set lines about ten miles off shore, were brought in.

SISKOWIT LAKE

Our next destination was Siskowit Lake, which is about nine miles long and one and one-half miles wide and is the largest lake on the island.

Mr. Fox, Dr. LaForge and Charles Purdy, crossed the three-mile trail to the lake and explored one side of it; while Professor Brownell and the writer, under the guidance of John Linklater, who carried a canoe to the lake, followed its other shore over another trail.

Upon embarking on the lake, we discovered a young bull moose swimming about a mile ahead of us. This was finally overtaken and photographed (Plate XIII, fig. 1). Professor Brownell insisted on riding the moose ashore, but our guide demurred so strongly that the idea was abandoned.

At the base of an up-turned tree on a sandy point, the writer found a perfectly formed obsidian arrow-head. At the same place, Professor Brownell secured a worked piece of rock, which appeared to be an unfinished pipe bowl. A dense jungle came almost to the water's edge, covering what appeared to have been an ancient camping-ground. Late in the afternoon, both parties reached the upper end of the lake in search of a village site said to exist there. The rocky location of this supposed site was so covered with moss that a careful investigation could not be made by us in the time allotted, as it was already 6:00 o'clock in the evening and we had nine miles of paddling and portage before us.

CHIPPEWA HARBOR

Our next night was spent at Chippewa Harbor, which has a very narrow and dangerous entrance. This little cove is hidden from the lake by high and rocky bluffs. Two fishermen with their families reside there during the summer months. Here was found an extensive Indian village site, which we excavated to quite an extent, finding potsherds, chert chips, fireplace stones, and various artifacts (Plate XIII, fig. 2). A grooved stone axe and a couple of arrow points are reported as having previously been found here by fishermen. Back of the harbor and on higher land, we found several elevations that appeared to be graves, but on excavation nothing was found in them.

Near the harbor entrance is to be seen an unusual log house, being two stories in height instead of one (Plate XIV, fig. 1).

BIRCH ISLAND SITE

This small island is located at the entrance of McCargoe Cove and is the summer home of Mr. John Linklater. At one end of this island is a village site which was examined by the members of the expedition with slight results. Mr. Linklater presented us with a broken arrow-point of black chert, picked up on his property. He, likewise, found a stone axe there and a white flint knife on the summit of a ridge near Sargent Lake. Since our return home, he has done some digging on this island and reports finding a few fragments of pottery and a number of flint flakes (Plate XIV, fig. 2).

Mr. Linklater, (Plate XV, fig. 1), whose ancestors lived on the mainland, informed us that his wife's grandmother and that his own grandfather remembered coming to Isle Royale,—that to appease the evil spirit supposed to dwell there, ceremonial dances were held on the Canadian shore before embarking in their canoes. They also told Mr. Linklater that when at the island they camped on its inland lakes rather than on the coast. This probably occurred not to exceed one hundred years ago and raises the question whether the village sites with their potsherds and stone artifacts were not the temporary homes of such visiting Indians.

The superstition existing in the minds of the relatives of Mr. Linklater seems to have been a survival of the one related by La

Ronde, to whom while at LaPointe in 1727 the Indians brought reports of a floating island of copper, probably Michipicoten, which no mortals could approach, since it was guarded by spirits who would strike any intruder dead.[4]

McCARGOE COVE

While the white yachts were riding at anchor in the quiet waters of McCargoe Cove, (Plate XV, fig. 2), Mr. Fox and Prof. Brownell explored Chickenbone, Livermore and Sargent lakes, where they located several probable village sites.

A walking survey of the mining operations back of the cove was made by Mr. Massee and the writer. Moose trails were followed for fully five miles through a dense forest of balsam, birch and spruce. Underbrush, root tangles, rocks and ridges of rock were encountered. After passing the workings of white men, who operated the Minong Mine for about nine years before abandoning it, we reached the most extensive district of prehistoric copper mining in not only the island, but in all America. Here, ancient workings almost touch each other for miles. It is the land of ten thousand pits and trenches, explored by a Milwaukee Public Museum expedition during the year 1924, of which the writer was a member. Mr. Massee stopped short and said, "When did this race of craftsmen lay down their tools and depart, who were they, from whence did they come and wither did they go?" My answer was, "We know not. Here is the open door, science is welcome to enter with its traditional tools, the lowly pick and shovel, to explore at will and this work seems to be for us".

We climbed a barren, rocky headland and were greeted by a most wonderful panoramic view, glorified by the setting sun. Far across the icy waters of Lake Superior and Thunder Bay lay the Sleeping Giant, a mass of rock towering twelve hundred feet above this great inland sea, marvelously carved in the figure of an Indian chief, with arms folded placidly across his breast.

The following day was spent by the members of the expedition in an archeological reconnaissance of this territory, and in making plans for its future exploration. Occasion was taken to trench an artificial mound, located by the Museum expedition of 1924 and which Dr. Barrett and the writer then thought might possibly be

[4]LaRonde, Louis, Wis. Hist. Coll., Vol. 17, p. 86.

prehistoric. It was found to be a dump of mine refuse, carried a considerable distance from a white-man's excavation. From part of an old trestle found within it, it is evident that an overhead carrier was used in making the deposit. Its top was covered by some two feet of black soil which had led to the suspicion that it might be a burial mound.

The expedition of the Milwaukee Public Museum in 1924 spent several days at McCargoe Cove for the purpose of studying the ancient mining operations that data might be secured for the construction of an environmental group for educational purposes. This group depicting an excavated pit of average size and showing the methods of aboriginal copper mining, is now on exhibition and has attracted much attention.[5]

While they were at the cove, a succession of pits and trenches were traced along the slope of the amygdaloid ridge for nearly five miles with a width of about four hundred yards thus covering an aggregate of one square mile. These pits when excavated are from four to ten feet deep and in some cases thirty feet across, usually cut into solid rock for a considerable distance (Plate XVI, fig. 1). The number was estimated at fully ten thousand. In those excavated, much charcoal and many broken mauls were encountered, fifty-six being taken from a single pit. These and other observations clearly indicated that the primitive method of mining by the use of fire, water and pounding was employed.

The mauls and hammer stones weigh from one to forty pounds, are usually oval in shape and are merely unworked cobble stones of granite or other hard rock. None of them appear to have been supplied with a handle or helve;—not even those recovered from pits, where they had lain undisturbed since being discarded by the ancient miners, showed the slightest evidence of such an attachment, which leads to the belief that the maul was held between the hands when in use (Plate XVI, fig. 2).

A Mr. Hart, who mined on Isle Royale during its boom period of mining, in speaking of the mauls so plentifully found there claimed that the greater part of them had grooves about them. Mr. Alfred Merritt, who was also there about the same time and who later for many years spent his summers at Toban's Harbor, dis-

[5]Barrett, Dr. S. A., Aboriginal Copper Mining at McCargo Cove, Isle Royale; Yearbook, Publ. Mus., Milwaukee, Vol. IV., pp. 20-27.

agrees with Mr. Hart on this matter and expressed his belief that the mauls were used without a handle.[6]

The belief expressed in Dr. Barrett's report that the thousands of mauls covering the ground in this vicinity were not necessarily brought from the mainland, as had been supposed, but were taken from ancient beach line and glacial deposits on the island, has been conclusively ratified by the present expedition.[7]

"Both Mr. Merritt and Mr. Hart have seen the cedar shovels or paddles found in the depths of some of the workings. These invariably shrivel and decay on being exposed to the air. They have seen the remains of wooden bowls, and Mr. Hart has in his possession birch bark taken from a pit and supposedly part of a basket or bucket used in removing earth or in bringing water. He also had a piece of rawhide thong found in the old workings."[8]

During the 1924 expedition, many excellent moving pictures of moose were taken at the head of the cove, where was then located a fine moose wallow. Owing to the waters of Lake Superior being much higher than they were at that time, this wallow was overflowed and but a few moose were seen by us at this place.

The number of moose on the island are estimated at from one to three thousand. Mr. Frank M. Warren of Minneapolis, who frequently visits the island, writes that he is of the opinion that fewer moose are there now than there were four years ago,—believing that the brush wolves (coyotes), which are increasing in numbers, are killing the calves. Mr. Charles C. Adams in listing the mammals of the island, as of 1907, makes no mention of moose.[9] Summer residents of the island all agree that moose were first seen there about fifteen to twenty years ago. The method by which the moose and caribou came to this lake-bound island must, of course, remain largely conjectural, but it seems probable that they came over the ice.

Although the work during the day was unusually strenuous, our evenings were made most profitable and entertaining, both of the yachts being provided with victrolas and moving picture projectors; the "Naroca" showing, among other things, very interesting reels of Arctic scenes; the "Margo", the sinking of ships

---

[6] Fox, Geo. R., Ancient Copper Workings on Isle Royale; Wis. Archeologist, Vol. 10, No. 2, p. 86.
[7] Barrett, Dr. S. A., Year book, Publ. Mus., Milwaukee, Vol. IV., p. 29.
[8] Fox, Geo. R., Wis. Archeologist, Vol. 10, No. 2, p. 91.
[9] Adams, Charles C., An Ecological Survey of Isle Royale, Lake Superior, Mich. Geol. Survey, Rept. 1905, p. 390.

during the war by German submarines. The "Naroca" had aboard a short wave broadcasting outfit and receiving set which kept us in touch with the outside world, enabled us to receive market quotations and to send reports to the press and other messages. The "Naroca" has the honor of sending the first message from the island over the air.

ISLE ROYALE POTTERY

The 1928 expedition to Isle Royale discovered several village sites or camping places, from two of which a number of small potsherds were obtained. One of these sites is located at Chippewa Harbor, and the other at Birch Island, near the entrance of McCargoe Cove.

The sherds were nearly a foot under ground,—none appearing on the surface. These are the first village sites discovered on the island and the first sherds secured there, so far as reported.

Considering these finds of grave importance, the fragments of pottery were turned over to Mr. W. C. McKern, Associate Curator of Anthropology at the Milwaukee Public Museum, for his opinion as to classification. Mr. McKern reported as follows:

"The potsherds from Isle Royale, which you have kindly made available to me for study purposes, are too fragmentary to permit of absolute classification. The shape of vessels which they represent, an important element in any pottery classification, is not apparent from sherds of this size, and the nature and placing of design elements are only indicated.

"However, the sherds conform in all general peculiarities to the Effigy Mound type of pottery. This differs from the pottery of three other known cultures of prehistoric Wisconsin in that it is lightly fired, grit-tempered ware, decorated primarily by means of cord imprinting, and in instances by means of narrow, incised lines. The most common ornamentation is a general surface roughening effected with a cord-wrapped paddle, with cord-imprinted or incised designs frequently added to the outer rim and lip of the vessel. Typical sherds showing this treatment are illustrated in plate XVII by the group marked 3.

"The base of a vessel of this type is not infrequently devoid of any surface marking, and this undecorated area in instances covers nearly the entire vessel. Undecorated sherds of this character are illustrated in plate XVII by the group marked 2.

"A second variety, or sub-variety of Effigy Mound pottery, of relatively rare occurrence, has no decoration other than a few widely dispersed and somewhat disorganized incised markings about the upper portion of the vessel. This type is illustrated in plate XVII by the group marked 1.

"Unfortunately, as you are well aware, no one has definitely succeeded in determining the types of pottery respectively made and used by the various historic tribes of Wisconsin. In so far as we have any data bearing on this problem, the Menominee seem to have manufactured an Algonquian type of pottery; the Winnebago, a distinctly different type, probably Siouan in character. In this connection, if the Effigy Mound type of pottery, to which the Isle Royale sherds seem to belong, was to be found anywhere in the eastern woodland area of North America, it would immediately and without hesitation be classified as Algonquian."

As the village sites discovered were all at a distance of two or more miles from any of the groups of aboriginal mines, and because of their limited size and the scarcity of the material they produced, it seems safe to conclude that they were merely temporary camping places, occupied for short periods of time by historic Indians, probably Chippewas, who occasionally visited the island, as we know they did, for the purpose of fishing.

Present-day archeologists are of the belief that pottery is the best index to the culture of the prehistoric inhabitants of this country. It is therefore hoped that the search for village sites, near to the ancient mines, may be continued and that the forest jungle of Isle Royale may cease to conceal the evidence needed to determine the culture relations of the miners.

ISLE ROYALE STONE AND COPPER ARTIFACTS

Aside from the thousands of stone mauls and hammerstones found about the numerous aboriginal mining pits on the island, a few other articles of utility in stone and copper have been recovered (Plate XVIII).

Figure 1 is a beautiful obsidian arrow point, one and three-quarters inches long, very nearly transparent, containing a dark longitudinal line almost its entire length and a concave base. This specimen was found by the writer at the foot of an upturned tree on the east side of Siskowit Lake.

Figure 2 shows a perfectly shaped arrow point of translucent chalcedony, one and one-half inches in length. This beautiful specimen was secured by Mr. B. A. Massee while examining the remains of an ancient dump of rock, from which the copper had been separated, located within the walls of Ferguson's "Old Fort", near Hay Bay.

Figure 3 is a finely formed chert knife, light drab in color, three inches in length, secured by Mr. Geo. R. Fox from the ossuary deposit, explored by the expedition at Point Houghton.

Figure 4 is a copper knife or spear point three and one-half inches long, found by the writer in a crevice on the crest of the ridge at McCargoe Cove, the eastern slope of which is covered with thousands of prehistoric workings. At Sault Ste. Marie, Ontario, we found in Mr. Turran's collection, three copper implements, two of which were of this type and found in the vicinity.

Figures 5 and 6 are small scrapers of chert, found on the village site at Chippewa Harbor. Each piece has a scraping edge, nicely formed by chipping. A number of chert flakes were also uncovered here. From the village site on Birch Island were obtained some flakes and a broken arrow point.

Mr. Frank M. Warren of Minneapolis, kindly furnished the photograph of five specimens secured by him on Isle Royale (Plate XIX, fig. 1). One is a finely made stone knife of white chert, found about two and one-half miles northeast of McCargoe Cove, lying on the bare rock. It is doubtless the one found by Mr. Linklater near Sargent Lake. One of the arrow points came from the north side of Rock Harbor and another from Birch Island, McCargoe Cove.

Mr. Frank M. Warren, of Minneapolis, reports by recent letter to the writer that in 1908 he recovered from the beach on the southerly side of Mott Island, at Rock Harbor, a copper arrowhead.

The finding of copper implements on the island by the early white miners and prospectors in excavating the ancient pits has been reported. Doubtless some were recovered by them, but no authentic record has been preserved and no such implements are now in existence, so far as the writer is able to find.

Ferguson reports the finding in 1922 of a highly polished axe head of diabase upon a rock overhanging one of the trenches. He found no other artifacts.

A recent letter from Mr. Emmet Scott of La Porte, Indiana, states that his father, while engaged in mining on Isle Royale in the seventies, secured at McCargoe Cove an excellently made copper spear-head which his father used for years as a letter opener, and a copper knife. After his father's death the spear-head became the property of his daughter, Mrs. Edward A. Rumley of 425 Riverside Drive, New York, N. Y.

Mr. Hart also reported the finding of a cache of several copper implements beneath a tree near the mines, probably at McCargoe Cove, which seems doubtful to the writer.

The ossuary deposits obtained from the "Massee Rock Shelter" were delivered to the Department of Anthropology of the Milwaukee Public Museum and examined by Mr. W. C. McKern, Associate Curator of that department, who reports as follows:

"An examination of the skeletal materials from Isle Royale, which has been placed with us for study, has been productive of the following data and observations.

"There are skeletal parts representing a minimum of twelve individuals. Of these, at least one was a child seven or eight years of age. Skull parts suggest the presence of at least three adult females. There appears to have been a normal variance in the physical dimensions of the adults; no exceptionally large or small individuals are indicated.

"The cranial specimens, in so far as they are sufficiently complete to permit measurements, are dolichocephalic (long-headed) and show not the slightest indication of occipital deformation due to cradle-board pressure. In this connection, the local historic Indians, according to available evidence, were unanimous in the use of the cradle-board, but practically all the crania found in Wisconsin mound burials, are free from cradle-board deformation.

"Of eighteen humeri, three have perforate olecranal fossae. All of the tibiae (shin bones) are markedly platycnemic, or sharply elongated in cross-section. This peculiarity is characteristic of the American Indians, although not uniquely so.

"Several of the teeth and associated maxillary bones show a diseased condition such as might result from abscesses. One femur exhibits boney excrescences on the anterior aspect in the region just below the greater trochanter. This condition may have been brought about by periostitis, osteomyelites or ostites.

"A more detailed treatment of the materials follows:

Osteological Specimens from Isle Royale

SKULLS

1 cranium—facial bones, maxillae and ramus missing.
maximum skull length—184.5 mm.
maximum skull breadth—132.5 mm.
cephalic index—71.81.
age—middle adult.
sex—possibly female.
Prominent occiput shows no evidence of cradle-board or other deformation.

1 skull—mandible and lower portion of cranium missing.
apparently dolichocephalic although measurements are impossible.
sex—female.
age—middle adult (sutures beginning to close.)

1 portion of cranium (portion of left parietal and temporal).
sex—apparently female.

1 portion of skull (frontal bone and mandible with right ramus missing).
age—child 7-8 years old.

1 right half of mandible showing evidence of abscess in region of second bicuspid.
age—adult.

1 front portion of maxilla, with bone necrosed in region of left first bicuspid evidently due to abscess.
age—adult.

CLAVICLES

2 clavicles.

SACRA

1 sacrum, complete, probably female.
1 sacrum, nearly complete, male.
1 sacrum, nearly complete, indeterminate.
1 base of sacrum, indeterminate.

STERNUM

1 upper portion of body of sternum.

VERTEBRAE

4 lumbar vertebrae.
2 thoracic vertebrae.
1 cervical vertebra.

PELVIC BONES

17 iliac portions.

SCAPULAE

3 right scapulae.
1 left scapula.

LEG BONES

Femurs
20 in number—maximum length, 47.2 cm. Adults.
1 in number—probable length, 22 cm. Child.
Fibulae
12 in number—maximum length, 35.5 cm. Adults.
Tibiae
24 in number—maximum length, 37.5 cm. Adults.

ARM BONES

Ulnae
7 in number—maximum length, 26.5 cm. Adults.
Radii
9 in number—maximum length, 26.5 cm. Adults.
Humeri
22 in number—maximum length, 33.3 cm. Adults.

NATIONAL PARK

As we delved into its mysteries, the conviction that Isle Royale deserved to become a National Park took a strong hold on the members of this expedition. As a result the following radiogram was sent to President Coolidge:

"As members of the Isle Royale Archeological Expedition and as Americans living in the Middle West, we suggest that federal action be taken in co-operation with the State of Michigan to make Isle Royale a national park or monument to preserve forever this northern land of woods and lovely waters for the people of this country. What scientists call "the riddle of the North" with its ten thousand ancient copper mines sunk into the rock before white man came, with its untouched wilderness of evergreens and birches and its scores of inland lakes, with its rugged shores and its bewilderingly beautiful estuaries reaching deeply into the interior of the island, as in no other place in the United States, with its mountains and its wild life and its lowlands, where thousands of moose are found tamer and more plentiful than anywhere in this country,—all this be preserved inviolable for the present and future generations of America and of the world. It should be preserved in the interests of archeological and geological science. It should be preserved for its natural grandeur and the beauty of its waters. In these fields it is unique. Though the northeasternmost island of the United States proper, Isle Royale is easily accessible by steamship from Buffalo, Cleveland, Detroit, Chicago, Houghton and Duluth, and by rail and steamship from all parts of the continent. Its profound archeological, geological and historical interest, its unique scenic beauty, its unparalleled wild life, its value as a health refuge from the heats and fevers of the country, and its accessibility to millions of people give weight to our suggestion. We submit this proposal respectfully to the President and the people of the United States."

A courteous reply was received by Commander McDonald, from the Secretary of the Interior, Mr. Roy O. West. It is also understood that as a measure of co-operation, an official inspection was made by the Director of the National Park Service, Mr. Stephen T. Mather, who spoke highly of the scenic beauty of the island and its surroundings.

To secure the desired result, it seems necessary that a new and determined start be made with a strong organization and competent leadership; that the State of Michigan, the various Isaac Walton clubs and other friends of the project co-operate. May the powerful influence of the press be exercised to this end.

Part II

PREHISTORIC COPPER MINING

THE SOUTH SHORE GROUP

Along the south shore of Lake Superior, ancient mining pits and trenches extend over a stretch of country known as the Trap Range, through Keweenaw, Houghton and Ontonagon counties. This range with a length of about one hundred and sixteen miles, and a width varying from two to five miles, runs almost parallel to the northern shore of Keweenaw Point (Map 2).

The first actual mining operations, within historic times, were commenced near the forks of the Ontonagon by Alexander Henry in 1761, but proved a failure. It was not until 1841 that the mining of copper again attracted attention. About this time, Dr. Douglas Houghton made a geological report to the legislature of Michigan, in which the earliest definite information in regard to the occurrence of native copper in the Lake Superior District was given to the public. Within a short time prospective miners and explorers flocked to this location. In 1846 the climax came and the resulting reaction forced all but five or six companies, out of the dozens organized, to suspend operations.

While the mineral belt was more or less pitted from one end to the other by the prehistoric miners, it was found that the most profitable deposits were located in three different groups; one on the waters of Eagle River, another at Portage Lake and the third near the forks of the Ontonagon River. In the bed of this river a chunk of mass copper, weighing nearly six thousand pounds, was found. The Indians had chipped away or cut off considerable of its bulk, and from this source and the metal obtained from nearby cliffs spread the fame of mines of copper among the ancient traders and missionaries. This mass of the red metal finally found its resting place in the National Museum at Washington.

In the bottom of the pits, which sometimes were from twenty to thirty feet in depth, it was not uncommon to encounter blocks or masses of copper weighing hundreds and a few of them thousands of pounds.

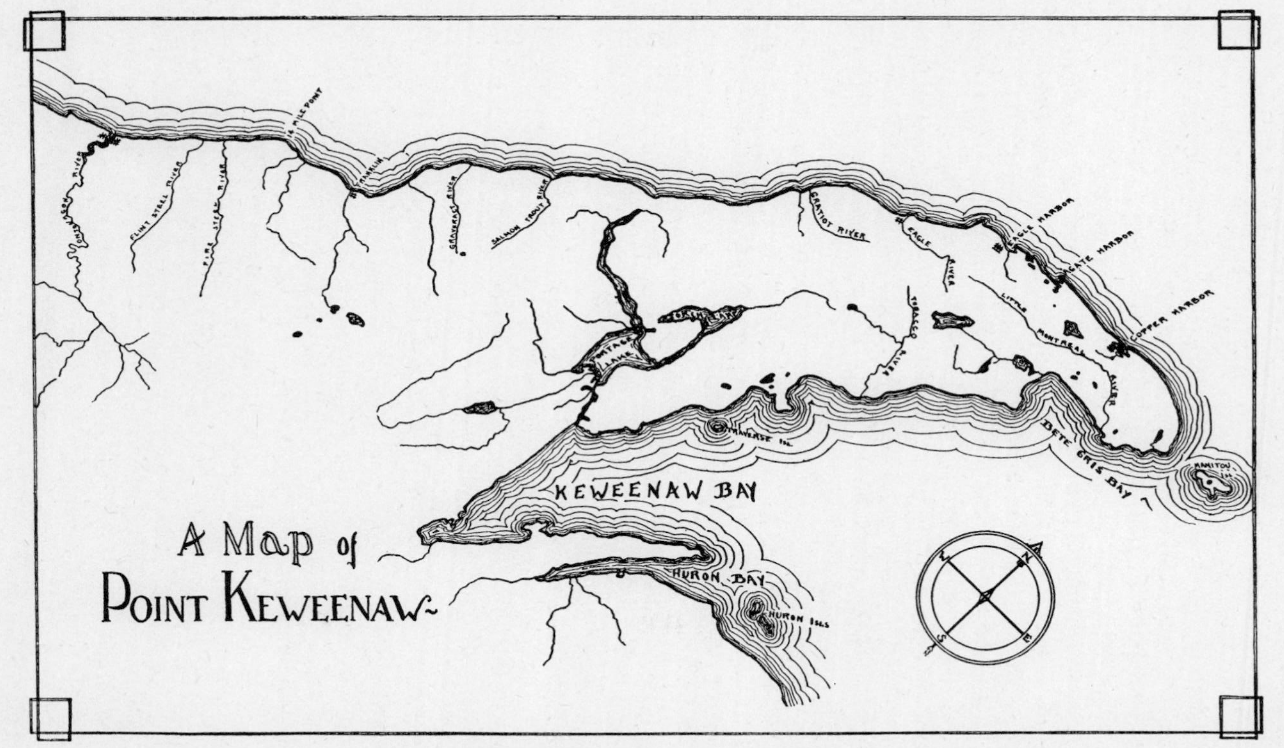

Map 2—Keweenaw Point, Lake Superior.

As early as 1875, when the writer conducted investigations along this copper belt, practically every pit had disappeared or become unrecognizable, as the result of exploration by white men. Thousands of mauls were in evidence of the same form and character as those found on Isle Royale. An occasional one, however, was reported to have a groove around the middle. In one instance a maul was secured weighing thirty-five pounds, and having two grooves for the attachment of handles. It was probably used by two men at the same time.

Aside from small workings near Marquette, Michigan, this location and Isle Royale produced all the copper used in the fabrication of the thousands of implements and ornaments found throughout Wisconsin and adjoining states. The successive glacial periods distributed detached pieces of this metal, known as float or drift copper, over an area measuring, according to Salisbury, about six hundred miles north and south, and seven hundred miles east and west, with the Lake Superior copper region at its northern edge.[10]

Wooden bowls for baling out the water, remains of bark baskets, used in removing the loose rock and dirt, and portions of timber thought to have been utilized as skids and ladders have been found in the pits among the rubbish. Some copper chisels and wedges have also been recovered. Paddles or shovels, of white cedar, resembling those now in use by the Chippewa for propelling their canoes, are reported by Whittlesey[11] as having been discovered in the pits. All of the mechanical contrivances employed were of a very simple nature, and the workings themselves illustrate nothing more than the endurance and patience of the Indian miners in their endeavors to possess the ore. With levers and men to use them the elevating of even the largest masses of copper which might be disengaged, could be accomplished without the application of any princple not understood by the most savage tribes.

METHODS OF MINING

Various descriptions of the methods of aboriginal copper mining in the Lake Superior Region have been published. Among

[10]Salisbury, Roland D., Notes on the Dispersion of Drift-Copper. Trans. Wis. Acad. Sci., Vol. VI, p. 42.

[11]Whittlesey, Charles, Ancient Mining on the Shores of Lake Superior, Smithsonian Inst., Cont. to Knowledge, Vol. 13, p. 8, 1862.

Hoy, Dr. R., Who Made the Ancient Copper Implements? Priv. Publ., Racine, Wis., 1886, p. 8.

Smith, Hamilton L., Annals of Science, Vol. 1, pp. 28-30, 1853.

these are a number in which the authors have carried to a ridiculous extent their fanciful conceptions of the engineering skill exercised by the native in obtaining the metal. As a fair sample of some of the exaggerated statements contained in these reports we may quote Schoolcraft who refers to these workings as:

"Vestiges of ancient mines so important in character that modern miners have paused in astonishment to behold".[12]

This statement is mild when compared to the wild guesses and conclusions of a score of other writers, most of whom had evidently never visited the region.

The following extract from a report of Dr. W. H. Holmes, then chief of the Bureau of American Ethnology, who made a careful investigation of the district, removes all doubt as to the character of the workings and of the methods employed by the Indian miners:

"The Lake Superior copper occurs in veins, bounded on either side by the hard metamorphic rocks making the upper peninsula of Michigan. The action of the atmosphere and of the acids from decaying vegetation upon the mineral, having produced a partial disintegration of the gangue, or rock in which it is held, the glacier scooped out deep troughs or channels in the rock thus softened. Often these depressions were only partially filled with drift, leaving more or less of the copper-bearing rock exposed as a wall on either side.

"Aboriginal mining in this region had its beginning in the hammering or cutting off of portions of the metal thus left visible; when the level of the gravel was reached, it was cleared away to follow the wall downward. From this it was but a step to removing the loose material in order to reach the copper vein at the bottom; and soon it was discovered that wherever one of these partially filled trenches occurred, copper was to be found beneath the gravel, whether any of it could be seen on the surface or not.

"When quarrying into the solid rock began, it was carried on in the ordinary Indian fashion, namely, by heating the rock, pouring water on it, and breaking up the fragments thus obtained, with stone hammers; perhaps using these hammers before the application of fire, so long as effective work in this manner was feasible or profitable. The hammers were rounded water-worn boulders, carried up from the lake shore or from the lower valleys. Modern

[12]Schoolcraft, Henry R., History of Indian Tribes, Vol. 5, p. 395.

work has shown that some excavations thus made were fully twenty feet in depth; and it is quite possible that others which have not yet been cleared out are much deeper".[13]

The American Indians are known to have mined deposits of other materials, which required fully as much toil as did the securing of copper at Lake Superior. The catlinite quarries of Minnesota, were worked until within a generation by the use of mauls and levers similar to those found in the copper pits. Mica was mined in North Carolina for a long period of time, and according to Prof. Kerr, the workings were not abandoned until after the advent of the whites. These mines often contained: "Open pits forty to fifty feet wide, by seventy-five to one hundred feet long and fifteen to twenty feet deep, and sometimes have tunnels fully one hundred feet in length".[14] These excavations extend all through Mitchell and the adjoining counties.

Dr. Holmes has investigated a large number of steatite and quartzite quarries, located in the valleys of the James, Potomac and Susquehanna rivers and describes their extent as "almost beyond conception." One of the best known and most extensive aboriginal flint quarries in the United States, is located at Flint Ridge, Licking County, Ohio, and upon which, judging from the descriptions of Fowke and other writers, an even greater amount of labor has been expended by the Indians than upon the workings at Lake Superior. The obsidian quarries of Yellowstone Park, the novaculite workings of Arkansas, and numerous others of lesser magnitude may all be cited, as evidences of hard and patient toil by the Indians, under press of necessity.

SOURCE AND USE OF THE METAL KNOWN TO EARLY HISTORIC INDIANS

Early European explorers and missionaries found copper implements and ornaments, or pieces of the metal itself, in the possession of many of the Indian tribes of eastern North America, from the St. Lawrence River on the north to the Gulf of Mexico on the south. The artifacts and metal so observed, together with the large number which have since been recovered from the mounds, graves

[13]Fowke, Gerard, Archeological History of Ohio, pp. 710-11.
[14]Kerr, W. C., Report of the Geol. Survey of North Carolina, Vol. I, p. 300, 1875.

and village sites of the eastern United States, investigation has shown, must have been procured from the Lake Superior country. Smaller quantities of the metal were probably obtained from places where copper occurs in Virginia, North Carolina, Tennessee, New Mexico, Arizona and elsewhere. The British possessions have deposits of this metal, but there is little reason to believe that the Indians utilized these sources of supply to any considerable extent.

It is also a well-known fact that at a very early date many ships came to the mouth of the St. Lawrence loaded with sheet copper and other metals to exchange for furs.

The early explorers in Canada, on the coast of New England, New York, Virginia, the Carolinas and Florida, including, besides Cartier,—Alfonse, Verrazano, Raleigh, Heriot, Ribault, Newport, Allouez, De Soto and Champlain,—all concur in saying that the Indians were using implements and ornaments of copper.

Rev. W. M. Beauchamp gives additional data bearing on the subject:

"When Bartholemew Gosnold was at Cape Cod in 1602 he saw a young Indian with plates of copper hanging to his ears. These may have come from European contact, but Gosnold did not suggest this. Farther south they were visited by natives, one of whom wore a copper plate, a foot long and half as broad, on his breast".[15]

Soon after Quebec was founded Champlain mentioned a very handsome piece of pure copper given him by an Algonquian. It was a foot long. The great discoverer said, "He gave me to understand that there were large quantities where he had taken this, which was on the banks of a river, near a great lake. He said that they gathered it in lumps, and having melted it, spread it in sheets, smoothing it with stone."[16] Instead of melting it, they probably softened the metal by heating and dipping it in water.

In 1666, Father Claude Allouez, founder of the first Catholic mission in Wisconsin, made the following statement of the Algonquian superstition regarding copper: "One often finds at the bottom of the water pieces of pure copper, of ten or twenty pounds in weight. I have several times seen such pieces in the savages' hands; and since they are superstitious, they keep them as so many divinities, or as presents which the gods dwelling beneath the

---

[15]Beauchamp, William M., Metallic Implements, New York State Mus. Bull. 55.

[16]Champlain's Voyages, Otis', Prince Society ed., Boston, Vol. 2, p. 236.

water have given them, and on which their welfare is to depend. For this reason they preserve these pieces of copper, wrapped up among their most treasured possessions. Some have kept them more than fifty years; others have had them in their families from time immemorial and cherish them as household gods.'"[17]

In a narrative, Father Claude Dablon, after treating extensively the Indian legend concerning the deposit of the red metal on Michipicoten Island, expresses the belief that the greater part of the copper came from Minong and that he learned from the Indians of the existence of the metal in abundance at that place,—that "pieces of copper, mingled with the stones are found at the water's edge almost all around the island, especially on the south side; but principally in a certain inlet near the end facing the northeast, toward the offing, there are some very steep clay hills where are seen several strata or beds of red copper, one over another, separated by other strata of earth or of rocks. In the water is seen copper sand as it were; and from it may be dipped up with ladles grains as large as a nut, and other smaller ones reduced to sand.'"[18]

The place, last described, must have been McCargoe Cove, and although Dablon did not visit Isle Royale, he evidently obtained quite an accurate description of it from the Indians. Although his report describes the location of the deposits, he says nothing indicating that the Indians, from whom he received his information, knew anything of the aboriginal mines.

Cartier found the natives of the whole seaboard "sparingly in possession of the red metal[19]"—Verrazano, in 1524, saw along the Atlantic coast, "many plates of wrought copper which they (Indians) esteem more than gold".[20] Newport was told by Powhatan, in 1607, that the copper they had "was got in the bites of rocks between cliffs in certain veins, a great distance north".[21] Abbe Segart, a missionary to New France in about the year 1630, gave an account of the resources of the country in his "Grand Voyage du pays des Hurons", and mentions specimens of copper from the Lake Superior mines, which he says were "eighty or one hundred leagues distant from the country of the Hurons".

---

[17]Wis. Hist. Coll's., Vol. 16, pp. 31-2.
Jesuit Relations, Vol. 1, pp. 265-67.

[18]Jes. Relations, 1669-70, French Regime in Wisconsin, Wis. Hist. Coll's. Vol. 16, p. 75.

[19]Wis. Hist. Coll's., Vol. 8, pp. 158-62.

[20]The Voyage of John de Verrazano, Coll's., N. H. Hist., Soc., 2nd ser., Vol. 1, pp. 47-50, 1841.

[21]Pickering, Dr., Chronological History, p. 926.

The French word for mines includes deposits of metal, in any form, that have a potential value.

The early French explorers were informed by the Indians, even as far south as the Gulf of Mexico, that copper came from a distant country in the north, and this information was verified as these voyagers neared the region of the Great Lakes.

In an account of Cartier's second voyage, 1535, given by Hakluyt, it is stated of the Indians along the south shore of the Gulf of St. Lawrence, that they said to Cartier that:

"The way to Canada was toward the west, and that the north shore, before Canada was reached, was the beginning of Saguenay, and that thence cometh the red-copper of them named Cavgetdage[22]."

Chrysostom Veryst, O. S. F. informs us that a celebrated Ojibwa chief of the Crane clan, named Tagwagane, who used occasionally to reside on Chequamegon Bay, Wisconsin, had a copper plate, an heirloom handed down in his family from generation to generation, on which were rude indentations and hieroglyphics denoting the number of generations of that family which had passed away, since they first pitched their lodges there. From this original mode of reckoning time, Mr. W. W. Warren concludes that the ancestors of this family first came to La Pointe A. D. 1490.[23]

Radisson, who wintered among the Beouf (or Buffalo) band of Dakota in Minnesota, in 1661-62, is the only one of the early visitors to this region who makes reference to articles made of native copper. In his description of these Indians he says:

"Their ears are pierced in five places; the holes are so big that your little finger might pass through. They have yellow waire that they make with copper, made like a star or a half moon, and there hang it."[24]

Henry, informs us that when he visited Lake Superior, at the time of the French war, the Indians obtained copper there, "which they made into ornaments and implements".[25] This was likely float-copper recovered from the beaches or hacked from the well-known mass of native copper lying in the bed of Ontonagon River.[26]

William N. Rogers, for several years connected with the Indian

[22]Jes. Relations, Voyage 1535.
[23]Warren, W. W., Wis. Hist. Coll's., Vol. 13, p. 430.
[24]Raddison, Sieur, Raddison's Fourth Voyage, Jes. Relations, Wis. Hist. Coll's, Vol. 11, p. 86.
[25]Henry, Alexander, Henry's Travels, p. 195.
[26]Jes. Relations for 1666-67, Thwaites ed., Vol. 1, p. 267.

Agency at Keshena, Wisconsin, told Dr. Hoy that he frequently saw copper implements in the hands of the Chippewas and Winnebagos. Many of their pipes were ornamented with copper.[27]

Saterlee Clark, Indian agent for the Winnebagos from 1828 to 1830, in a personal communication to Dr. Hoy, said:

"When I first came among the Winnebagos many of them had copper-headed weapons. Many of them carried lances headed with copper."[28]

One celebrated calumet which formerly belonged to Black Hawk, and was later owned by a Winnebago chief, had a broad rim of copper with great blotches of native silver.[29]

In recent years copper implements have frequently been obtained from graves and burial mounds in various parts of Wisconsin, thought to be those of historic Indians. Hoy took from each of three Indian graves in the town of Caledonia, Racine County, which were side by side, several beads and cylindrical articles of copper three or four inches in length. In one of these graves were found "two blue cut glass beads". Nearby, another copper implement, together with some human bones, were secured by William Hess.[30]

In 1870 the writer procured from a small burial mound, Racine County, a copper spear point and two irregular pieces of copper showing hammer marks, together with a copper or brass kettle of European manufacture, and the remains of a skeleton.[31]

The historic evidence as to the use of copper by the Indians is conflicting, as shown by the following quotations:

Legler says, "in the written accounts of some of the early French voyaguers in this region, which have been preserved in the archives at Paris, there is no reference to the use of copper tools, though the knowledge of the metal, which the Indians of that period regarded as a sacred gift, not as an article of utility, is frequently mentioned".[32]

Robert de La Salle, who made a voyage from Green Bay to the site of the city of Milwaukee, about two hundred and fifty years ago reported, "the extremity of their arrows is armed with a sharp stone or the tooth of some animal, instead of iron. Their buffalo

[27] Hoy, Dr. P. R., Who Made the Ancient Copper Implements?, p. 1.
[28] Hoy, Dr. P. R., op. cit., p. 11.
[29] Hoy, Dr. P. R., op. cit., p. 11.
[30] Hoy, Dr. P. R., op. cit., p. 13.
[31] Hoy, Dr. P. R., Who Built the Mounds?, p. 26, priv. publ., Racine, 1886.
[32] Legler, Henry E., Leading Events of Wisconsin History, Chap 3, p. 18.

arrow is nothing else but a stone or bone, or sometimes a piece of very hard wood".[33]

Father Louis Hennepin tells in his history that the Indians instead of hatches and knives employed sharp stones and instead of awls used pointed bones.[34]

The French historian, Charlevoix, who visited Wisconsin in 1720, described Indian hatchets of flint, which he says were the only implements used in felling trees.[35]

Although several of the ancient Jesuits and traders, who first visited what is now Wisconsin, did not report the use of copper by the savages, it is quite certain some of these artifacts were in the possession of the savages at that time. That their records make no mention of the use of copper implements is not surprising, as no report was made by them of the existence of thousands of effigy and burial mounds among which some of the travelers are known to have camped. The occasional finds can well be considered as mere echoes of the thriving industry which lasted for many centuries and was at its peak long before the Discovery.

It is a well-known fact that the streams were the high-ways of the savages and that their village sites, because of a desirable location, were sometimes occupied by successive tribes, as the different cultures discovered indicate. That the Winnebagos, as well as other tribes of Wisconsin Indians, were well advanced in agriculture is evidenced by the several hundred corn fields and garden beds that have been discovered within the borders of this state and attributed to them.

The few copper artifacts of history may have been heirlooms, handed down from many generations, or they may have been recovered from village sites or from the corn fields and garden beds of their predecessors, and have been the means of misleading many students of archeology into the belief that copper mining by the Indians was carried on after the white-men came.

NO MINING DURING THE FRENCH REGIME

The French discoverers and explorers of this region were deeply interested in mining operations and revealed two great deposits of metal, copper and lead, in the Northwest. The difficulties of

[33]Wis. Hist. Coll's., Vol 7, p. 90.
[34]Wis. Hist. Coll's., Vol. 7, p. 90.
[35]Wis. Hist. Coll's., Vol. 7, p. 90.

transportation hindered them in utilizing these metals to any great extent and made early mining a promise rather than a performance.[36]

Although some of the French governors of Canada knew of the existence of the copper deposits and were very anxious to develop the resources of the country and make them valuable to the French king, all attempts at mining were merely feeble efforts. There is no historical authority, and no facts have yet been discovered about these ancient mines, to warrant the supposition that the French participated with the Indians in the process of copper mining.[37]

The following is a concise summary of the early history of copper mining in the Lake Superior District.

"The early fame of Ontonagon River was due to the copper found upon its banks; although the first known mention of the stream alludes to the large sturgeon fishery near its mouth. As early as 1665 reports of copper mines were sent out from Lake Superior by voyageurs and Jesuit missionaries. In 1668 a considerable nugget was sent first to the intendant Talon, and later to the king in France. Hence, on one of the earliest maps, the river is designated "Nantononagon or Talon", but the latter name soon disappeared. Aside from the nuggets of copper found, there was a large boulder of virgin copper lying upon the banks of the Ontonagon, some twenty-five miles above its mouth. This caused the French to believe that a copper mine might be discovered in the near vicinity. In 1735 Denis de la Ronde, then commandant at Fort Chequamegon, asked the French government for experts to aid in locating these mines.[38] Three years later, a German miner, named John Adam Forster, and his son, explored this vicinity at the instance of La Ronde, and made favorable reports thereon.[39] But a fierce Indian war and the subsequent death of La Ronde, ended the mining projects of the French in the Lake Superior district. The earliest English attempt was that of Alexander Henry and his partners in 1772.[40] Douglass Houghton, in his famous geological report of 1841, alludes to this effort, and the lack of scientific knowledge shown in making locations. From Henry's time

[36] Kellogg, Louise P., French Regime in Wisconsin and the Northwest, p. 363.

[37] Winchell, N. H., Aborigines of Minnesota, Minn. Hist. Soc., p. 502.

[38] Thwaites, Dr. Reuben Gold, Wis. Hist., Coll's., Vol. 17, pp. 237-240.

[39] Ibid., pp. 306-315.

[40] Henry Alexander, Travels and Adventures in Canada and the Indian Territories, pp. 225-229, N. Y., 1809.

until the advent of Americans upon Lake Superior, no further effort was apparently made to explore for copper mines".[41]

RELATIVE AGE OF PITS

Condition of Wood Not Conclusive

The well-preserved condition of wooden implements found in the aboriginal pits caused early archeologists to believe these mines to be of no great age.

Dr. Jacobson of Boston, who spent several years on Lake Superior during the early period of copper excitement, told Dr. Hoy in 1844, that: "The fresh condition of the wood-work, skids and ladders, and the fact that sharp axes were used in fitting the timbers, is evidence that they were not of great antiquity."[42]

Dr. Lapham, who never visited the copper region, thought the sleepers, levers, wooden bowls, etc., to be "rather indicative of Caucasian ingenuity and art."[43]

The writer, during his visit to the mining region on the South Shore in 1875, saw a ladder, two paddles and fragments of a birch bark receptacle, taken from one of the ancient pits. These were all in a good state of preservation, excepting that the birch bark when dried was very brittle. This ladder, like many others that were found in that district, was merely a trunk of a small tree with limbs cut away, leaving stumps of about six inches in length. This is the well-known form of Indian ladder. The one seen was cut off by hacking, with some instrument, clear around the portion to be cut and not cutting from two sides, as a white man does. The instrument used was reasonably sharp and might have been a copper or stone axe.

An expedition of the Milwaukee Public Museum, in excavating the famous ruins at Aztalan, found in dumps and refuse, below the water level of Crawfish River, a number of posts that had certainly lain there for centuries, which were cut in the same manner and showed slight evidence of decay.

Some years since, the writer was privileged to visit a brick yard near Neenah, Wisconsin, where an acre or more of clay had been excavated to a depth of about twenty feet. Below this an ancient tamarack swamp was encountered. The trees were pros-

---

[41]Thwaites, Dr. Reuben Gold. Footnote, Wis. Hist. Coll's., Vol. 19, pp. 182-3.
[42]Hoy, Dr. P. R., "Who Made the Ancient Copper Implements?", p. 8.
[43]Lapham, Dr. Increase A., The Antiquities of Wisconsin, p. 76.

trate, all lying toward the southwest and well preserved. This interglacial forest had been buried by the deposits of red clay in one of the later stages of Wisconsin glaciation, thousands of years ago.

In drilling an artesian well in Racine County, Wisconsin, the trunk of a tree was struck, many feet below the surface and covered by two distinct glacial drifts. All of which proves conclusively that wood beneath water or wet mud will resist decay indefinitely and such implements found in aboriginal pits beneath the water level, as was the case with all the finds in the South Shore group, have no bearing on the question of age, so far as their state of preservation is concerned.

Age of Mines Cannot Be Determined By Trees

A familiar argument offered in favor of the great antiquity of the earthworks of our own and other states, as well as of the mines, is that of the large size of the trees and stumps frequently located on or near them. Such statements occur in many of the older works on American archaeology. Other later authors have shown that the age of large trees, is quite commonly overestimated.

The celebrated redwood stump of California, used as a floor for a ballroom, was determined by J. G. Lemmons, in 1875, to be just 1,260 years old, although oft reported at from 2,000 to 3,000 years.[44]

That the soil, location and climatic conditions influence the growth of trees is certain. In order to investigate this matter, the writer examined three white cedar trees each one foot in diameter, exclusive of the bark, and found them to contain forty-nine, sixty-four and eighty-one rings respectively. While on the last trip to Isle Royale, there was cut down a white birch, five inches in diameter, which contained fifty rings of growth. After returning, the writer cut one in upper Wisconsin, ten inches in diameter, and found it contained but twenty-one rings, indicating that the growth on the island is slower than it is two hundred miles farther south, probably because of climatic conditions and the shorter seasons in the North.

Fowke has shown that: "All these data are uncertain guides. The rubbish may have lain for a long time before the particular tree in question began to grow."[45]

[44]Hoy, Dr. P. R., Who Built the Mounds?, p. 14.
[45]Fowke, Gerard, Archaeological History of Ohio, p. 709.

Age of Trees—How Determined

In regard to determining the age of trees by counting the rings of growth, Dr. Thomas says:

"Recent investigations have served to destroy confidence in the hitherto supposed certain test of age, as it is found that even within the latitude of the northern half of the United States, from one to three rings are formed each year; and there is no certainty in this respect, even with the same species in the same latitude."[46]

Believing that Dr. Thomas' conclusions might not apply to the Lake Superior District, the matter was referred to Mr. Huron H. Smith, Curator of the Department of Botany, Milwaukee Public Museum, whose conclusions are as follows:

"In the Temperate and Arctic zones, all broad-leaved and evergreen trees add a distinct ring of annual growth, and their age may be definitely obtained by a careful count with a hand lens. In the tropics and subtropical zones, these rings are not so definite, although some species do add definite rings for one year of growth.

"The diameter of the tree is not always an indication of the age, however, for favorable conditions of growth may cause the annual rings to be very large, while in other trees that have had much competition for light, the rings may be very small. A cross section of the tree trunk reveals many interesting facts about the growth of the tree. To one skilled in interpreting these facts, it tells the date of an unfavorable season, as well as the favorable seasons. One can read from the rings, which side of the tree was most shaded, and can also see the tragedies of wounds from falling trees or from fires.

"In general, the farther north one goes, the smaller are the annual rings, even in the same species, which is another way of saying that growth is not so great at the northern limits of that particular tree. A white birch that would measure eighteen inches and show an age of sixty years in central Wisconsin, would compare with a five-inch trunk on the northern shores of Lake Superior, which would probably be the same age. Going farther north, it has been found at the limits of tree growth, in the 'land of little sticks', that sometimes a birch tree one hundred years old, will be only six feet high and have a trunk diameter of half an inch.

"Another example of the disparity of size and age is shown in

[46]Thomas, Cyrus. Report of Mound Explorations, Bureau of American Ethnology, Vol. 12, p. 627.

two trees I counted in Humboldt Co., California. These redwoods were eighteen feet nine inches in diameter and nine feet three inches in diameter, breast high. The larger one was one thousand, one hundred and seventy-nine years old and the smaller two thousand, two hundred and sixty-eight years old. The greatest care was needed to count the smaller one with a hand lens, placing pins at the end of every ten year period.

"All increment studies by the Forest Service are based upon the study of the annual rings, and mature timber is determined by the results of these studies, which are the basis of all sales in the national forests."

A recent letter from Raphael Zon, Director of the Lake States Forest Experiment Station, University Farm, St. Paul, Minnesota, confirms the information received from Mr. Smith, and adds the following:

"You, of course, know that to get the total age of the tree, the rings must be counted at the base of the tree or what we call the root collar; that means the line where the stem and the roots join. If you count the rings on stumps of different heights, you must allow a few years for the growth of the young trees from the ground to the particular height. For our ordinary stumps of a foot or eighteen inches in height, about five years should be allowed. Rings on any other section of the tree tell you only the age of the part above that section but not below it."

No careful study of the age of the forests found on Isle Royale seems to have been made. Trees of considerable size are frequently found growing within the pits and on the dumps but few of them could possibly be pre-Columbian. An older forest is in evidence, as stumps and prostrate trees of a much larger size indicate, but which are in such a state of decay as to make the determination of their age almost impossible (Plate XIX, fig. 2). The size or age of trees growing on the dumps are but limited factors in determining the antiquity of the aboriginal mines for the reason that the excavations may have been centuries old when the tree began its growth.

THE STORY OF PREHISTORIC COPPER MINING

The pits, the stone hammers, the charcoal and the artifacts made of copper, are the only relics, so far discovered, of the race that did the mining. Here is to be seen a most remarkable illustration of the ingenuity and practical skill, which man in a rude state of culture can summon to his aid according as his necessities demand.

Before proceeding to their summer's work in the copper region, considerable preparation was necessary in order that the necessities of the trip might be provided. A party making the pilgrimage had to be organized. Those chosen would naturally be the most stalwart of the village, doubtless including some of their women folks; the older members of the family remaining at home to look after the children and attend the garden beds. That the women were equally as able to paddle a canoe as the men is a historical fact and the division of labor among them was well defined.

In proceeding to Lake Superior, it is probable that they followed trails through the virgin forests for a hundred miles or more, packing with them their weapons, tools and other necessities. When they reached the great lake, those who failed to bring canoes were compelled to build boats in order to continue the trip to Isle Royale. They were obliged to follow the coast-line until opposite the island, making the distance very great and could cross only when the sea was calm. At the very best, this was a hazardous trip.

After reaching the island, the camp had to be prepared. They thoroughly combed the beaches for copper nuggets, but this source of the metal soon became depleted.

After the camps were established, the division of labor again asserted itself; some set to work removing the surface soil and moss from the rocks in search of the green streak of malachite, which they were to follow. Others were engaged in gathering the necessary fuel for mining purposes and as the available down timber was soon exhausted, the felling of large trees became necessary. This was accomplished by the slow process of building a fire around the base and chipping away the charred wood with copper or stone axes until the tree fell. It had then to be cut into sections, split into convenient sizes and transported to the prospective mine.

Others were obliged to procure such food as they could in addition to the jerked venison, and dried vegetables that were brought from their home. Game on the island soon became a scarcity. That fish, however, were plentiful and must have constituted one of their principal means of sustenance, is indicated by fish scales now found in a number of the pits.

It became necessary for others of the party to procure from the beaches and glacial deposits and to transport to the mine as fast as needed, water-worn boulders of diorite, granite, greenstone or porphyry, which were used by the miners as hammers and sledges.

Still others carried water in receptacles of birch bark or in wooden bowls brought with them or made as required.

All preparations being completed and the streak of green in the hard rock having been discovered, fire was started by the primitive method of drilling a laterally placed piece of hardwood with an upright stick of softer material, and applying tinder. When the rock was very hot, water was dashed upon it, which partly disintegrated it by contraction. The weakened portions were pounded with stone mauls to remove the particles of rock adhering to the copper. These mauls would weigh from five to forty pounds and were held between the two hands when in use. As they are very hard and brittle, their usefulness was of short duration since the ends were easily fractured. They were then discarded, which practice accounts for the hundreds of thousands of hammerstones now lying upon the ground and recovered from the pits.

This treatment was repeated over and over. Each time the depth was increased by a fraction of an inch until the nugget of copper, which underlay the green streak was recovered. In many cases their efforts were in vain. In order to gain the desired depth, the pits had to be widened and in many instances became trenches from five to fifteen feet in width, six to ten feet in depth and about twenty feet in length. In order to remove the debris, their paddles were used as shovels, some of which have been recovered from the pits.

The copper was separated from its matrix by the use of hammerstones handled with one hand and weighing from one to three pounds. The mauls and hammerstones recovered on the island were not supplied with handles. Their period of usefulness seems to have been too short to warrant helving them. On the mainland, however, a few grooved mauls have been reported, but the writer saw none among many hundreds that covered the ground in the vicinity of the pits on the South Shore.

The depth of the mines in either district was limited by the ability of the miners to bale out the water as they seemed to have no mechanical means of elevating it. Where the pits were of considerable depth and extended below the water line, they used the Indian ladder and baled the water so long as it could be successfully done.

As winter approached, the rigors of the climate caused mining to cease and those who survived, carrying with them the result of the summer's work again braved the turbulent seas in their frail crafts in order to rejoin their families.

HISTORY OF THE NUGGETS

The mode of fabrication of copper implements has elicited not a little discussion among archeologists. All evidence, as to the manner of manufacture, is derived from their superficial appearance and experiments by archeologists in their duplication. The history of the fabrication of pre-historic copper artifacts, previous to the coming of white man, is unrecorded and only inferentially known. It is, however, a well-established fact that each Indian village had its "arrow-head makers". These were old and incapacitated men, unfit for war or the chase. These ancient artisans became most proficient in their art. It seems reasonable to suppose that the copper nuggets were turned over to these men to be manufactured into such implements and ornaments, as were most in demand.

While historical reports state that the Indian reverenced certain chunks of copper as divinities, it is more than likely that these were the ones secured from the drift and not the product of the mines. This the writer judges from the fact that nuggets found in the drift are usually worn smooth and into fantastic shapes; while the copper from the mines is not of a shape that would work upon the imagination or awaken the superstition of the savage.

The ancient fabricators were not only skilful in the production of the artifacts, but adhered strictly to well-known types or forms. In the process of manufacture, the nuggets were assorted into groups that were best adapted for the making of specific types.

That the ancient copper-smith knew and practiced the arts of annealing and welding in working native copper is evidenced by many examples of his handicraft. He doubtless found, as have archeologists who experimented with this metal, that when the untreated copper was pounded to a thin edge it crumbled, but that by frequent heating and dipping in water during the fabrication of the implement it became tough and easily worked. This process is called annealing.

As a result of pounding, the cutting edges became hard, which led to the erroneous conclusion that the prehistorics understood the tempering of copper.

The artifacts contain a small amount of free silver, which could not be seen if the metal had been melted, and no implements of copper, yet recovered, were tempered. By means of the annealing process and careful hammering, the prehistoric was able to weld separate pieces of copper into a homogeneous mass. In all large collections of copper implements may be found examples indicating that portions of thin copper were folded back over the blade and welded to it, and in a few instances complete objects were formed by the welding together of two separate pieces of equal size.

Their anvils were smooth rocks. Their hammers, with which the nugget was beaten into shape, were also of rock,—the finishing was done with rubbing stones and a final polish given by the use of sand and water.

Copper chips of suitable size were often made into beads by a simple process of bending the fragment around a core, which with a small amount of hammering and polishing, produced the finished product.

Although objects made of sheet copper are rarely found in Wisconsin and adjoining states, the native copper-smith, with the use of the tools at his command, could easily have reduced the nugget to sheets as thin as he might desire by the processes of annealing, grinding, cutting, embossing, perforating, and polishing.[47]

In the production of sheet copper, the flattening of nuggets was not always necessary. Very thin sheets of copper are occasionally secured from seams in the copper-bearing rock of the Lake Superior District. The writer procured one specimen from the South Shore group, Upper Michigan, about three-sixteenths of an inch in thickness and from which could have been cut a rectangular piece, seven inches in width and ten inches in length. This specimen was for a considerable length of time in the collection of the late Mr. Charles Quarles, Milwaukee, Wisconsin. From this specimen, with slight

---

[47] Willoughby, Chas. C., Primitive Metal Working, Discussion As to Copper from the Mounds, Moore, McGuire, Putnam, Dorsey, Moorehead and Willoughby. p. 55.

grinding and polishing, a sheet of copper could have been produced undistinguishable from the European product, excepting by chemical analysis.

WHERE ARTIFACTS WERE MADE

No evidence has been found that artifacts of copper were manufactured on Isle Royale. More implements made of this metal have been found in Wisconsin than in all the other states of the Union combined. Minnesota,[48] although nearer to the mines, has yielded but a few hundred, Upper Michigan possibly a thousand, Canada but a small number while the fields, village sites, mounds and graves of Wisconsin are estimated by Mr. Charles E. Brown, Secretary of the Wisconsin Archaeological Society, as per a recent letter to the writer, to have produced fully 20,000 specimens. The bulk of these finds in Wisconsin were secured in the counties bordering Lake Michigan. When the forests of Upper Michigan, south of its copper mining area, are cleared away, it is reasonable to suppose that large numbers of copper artifacts will be found there.

Up to 1873 when the late F. S. Perkins began a drive for coppers, a considerable number of such implements must have been destroyed or otherwise lost. Many of them were reported as sold to junk dealers. The total number recovered to date is probably but a very small percentage of those manufactured. The writer estimates that the weight of all copper implements and ornaments, made of Lake Superior copper, thus far recovered, does not exceed two tons, which, compared to the great amount of work done at the mines, is insignificant.

Workshops or places where these artifacts were made have been located in many parts of Wisconsin.

The late H. P. Hamilton reported that immense quantities of flint chippings and pottery, as well as chips, scales and fragments of copper have been uncovered by the shifting sands on an extensive ancient village site at Two Rivers, Wisconsin, and the evidence appears to be conclusive that implements of copper as well as stone were manufactured there.[49]

At the Wisconsin State Fair in 1905, there was exhibited a portion of the collection of S. D. Mitchell, which contained a large number of copper chips taken from the village sites in Green Lake County.

---

[48]Winchell, N. H., Aborigines of Minnesota, Minn. Hist. Soc., p. 498.
[49]Hamilton, Henry P., Wisconsin Archeologist, Vol. 1, p. 11.

In 1910, the writer discovered in four places along the banks of the Fox River in Marquette County, large numbers of copper chips, these evidently being workshops or places where the copper implements were made. The chips found here, as in all cases, were merely scales produced by hammering and too small for practical use. Thousands of them must have entirely disappeared as the result of oxidation.

Another place, where copper artifacts were fabricated, was located on the Black River site, south of Sheboygan, Wisconsin.

Mr. Francis S. Dayton of New London, Wisconsin, recently reported to the writer that he had collected many copper chips and flakes from a number of village sites in Waupaca County, this state. His finds in copper have amounted to about a thousand, which include some finished implements and a number of partly completed ones. He also reports having secured drills, punches and other implements used in the manufacture of copper artifacts.

One of the punches in the collection of Mr. Dayton is unique in shape, being two inches long and one and one-fourth inches square at the top. Its head is much battered and its point is turned to one side, perhaps intentionally, so as to give the user an unobstructed view of his work.

Mr. J. P. Schumacher of Green Bay, Wisconsin, an archeologist of note, recently reported to the writer that:—

"All of the collectors here have found copper chips on the different sites in this region. Personally, I have found them on practically all of the sites along the lake-shore and on both sides of Green Bay. On the west side, on the Elmore property, I discovered a cache, some ten years ago, of one copper spear, one copper awl, and about half a pint of small copper chunks from the size of a bean to a good-sized marble. From the town of Pittsfield, I obtained a knife, hammered out of copper, which showed fifty or sixty specks of silver. I also have some large copper specimens, which show evidence of having been worked on. One of these was found in the town of Union, Door County, and one on a site near Garden Bay, Michigan."

These and several other evidences of workshops, within the state, together with a number of unfinished implements, have been called to the writer's attention.

Mr. J. T. Reeder of Houghton, Michigan, who has lived on the

Copper Range for more than forty years and is the greatest collector of copper implements in Upper Michigan, recently informed the writer that, in his opinion, not over seven hundred and fifty to one thousand copper artifacts had been found within a radius of fifty miles of his home,—that the greater number were found along the Portage Lake Canal, which is twenty-eight miles in length. Sloats Creek, Pilgrim River, Dollar Bay, the Messner Farm and other locations were also productive.

"From a cache on a farm near Osceola, about two and one-half miles from Calumet, I secured and still own one hundred and fifty pieces or more, eight or ten of them completed celts, several partly formed ones and many fragments. These fragments were small and rough, apparently as they were detached from the rock. This was evidently a regular workshop.

"Another workshop was found on the bank of Snake River, near Fighting or Battle Island. A third was located one and one-half miles southwest of Chassell, Michigan, where twenty or more celts, many partly formed objects and fragments were found."

Mr. Reeder is of the opinion that the miners from the south on their way to Isle Royale, passed through Portage Lake, returning the same way. At the Entry, the last stop before reaching Lake Superior, have been found many copper implements, among them several fine spuds. From other evidence, he believes that the copper was here freed from adhering fragments of rock and implements were fabricated. He does not believe that permanent camping sites ever existed along the shore of this lake, as no Indian cemeteries have been discovered there.

Many years ago the Milwaukee Public Museum purchased from Mrs. William Oppenheimer, the collection of Mr. and Mrs. Daniel Washburn which consisted of 397 specimens of copper, secured by them near Hancock, Michigan. It contained, among other things, 134 partly worked pieces and many fragments, indicating that they were mainly rejects from workshops.

Mr. Towne L. Miller, Honorary Curator of Archeology of the Milwaukee Public Museum, who visited the Museum of the University of Pennsylvania, at Philadelphia, and examined its collection of copper artifacts from Wisconsin, known as the Mitchell Collection, found among other things, 311 fragments of copper.

Mr. Miller reports that he has found camp sites in sections 7, 18 and 24 in the township of Marquette, Marquette County, Wisconsin, where he secured some copper fragments. The sand blows from which the copper was taken are scattered along the river bank for five miles with other camp sites on the opposite side of the stream. On one camp site, as the result of five visits, "I found 7 beads, 15 awls, 2 fish hooks and 49 fragments. This site is about 3 rods wide and 20 rods long. On another, containing an acre, in two visits, I secured 2 beads, 4 awls, and 49 fragments."

As nothing of this character has been reported from Canada or Minnesota,—all evidence leads to the conclusion that an area including Wisconsin and Upper Michigan was the home of the copper artisans and that the making of implements from the red metal was carried on almost exclusively here.

THE DISTRIBUTION

Wisconsin being the seat of this industry, distribution must have taken place principally from here. That trade existed with distant tribes is evident from the fact that conch shells from the seacoast are sometimes encountered in our mounds. Artifacts, made of obsidian, which was undoubtedly quarried in the Yellowstone country, are found in the mounds and graves, and on the Indian village sites of this state. Specimens of ivory-colored flint, probably quarried in Ohio, are quite common. Many of the pipes in the G. A. West collection at the Milwaukee Public Museum are exotics although Wisconsin finds and are made of material and in forms foreign to this state. These together with ornaments, ceremonials and other enduring objects that are found here, must have been acquired in exchange for something else, which it is reasonable to suppose was, principally, manufactured implements of copper. Thus, through channels of trade, which are known to have existed between the ancient tribes of America from time immemorial, the products of the skilled copper-smiths of Wisconsin passed from hand to hand and from tribe to tribe to far distant lands.

Although the copper artifacts of the south differ radically in type from those characteristic of Wisconsin, some of the raw material, from which they were made, came from the Lake Superior District.

Mr. Clarence B. Moore, who has done most valuable work in

mound exploration in Florida and other southern states, is authority for the statement: "So many evidences of prehistoric intercourse with regions to the south have been found in the mounds of our western states that it is safe to assume that the Lake Superior district furnished the greater part of the copper in use by Southern Indians, which was doubtless traded for shell implements and ornaments, or for the raw material obtainable only on the seaboard or the Gulf Coast."[50]

WHO WERE THE MINERS?

Just who were the primitive miners of Lake Superior is still a matter of conjecture. Students of archeology are almost unanimous in the belief that they were Indians and none but Indians. That they belong to one of the two great families, the Siouan or the Algonquian, seems to be quite certain. It has been suggested that this mining was done by no one tribe and that warring tribesmen forgot their hereditary hatreds while they dug the metal, thus making the district neutral ground, as was the red pipe stone quarry of Minnesota. This seems improbable, there being reason for the belief that the pre-Columbian copper mining of the upper lake region was done solely by some one people, strong enough to hold the same against other tribes. It is not reasonable to suppose that the possessors of the source of a metal so valuable for the manufacture of implements of war and the chase, would permit outsiders to enjoy the same advantage.

The original habitat of the Siouan peoples is one of the unsettled problems of American anthropology. A theory, proposed by Horatio Hale, that their original home was on the eastern slopes of the Appalachian Mountains, from which they slowly migrated to the Ohio Valley and for some reason later advanced both up and down the Mississippi and along the Missouri was accepted by Dorsey, Mooney, McGee, Brinton and Livingston Farrand.[51] A portion of them came into Wisconsin around the south shore of Lake Michigan and others followed the Rock, Wisconsin, and other rivers leading into the central parts of the state and finally reached Lake Winnebago, which was given the name of this branch of the Siouan family.

---

[50]Moore, Clarence B., Florida, p. 258.
Fowke, Gerard, Archaeological History of Ohio, p. 706.

[51]Dorsey, J. Owen, Siouan Sociology, Bur. Am. Ethnology, Fifteenth Ann. Rep.

Other investigators claim that the original Siouan habitat was north of Lake Superior, from which they advanced both east and west reaching into the Mississippi and Ohio Valleys.[52]

Regardless of their original habitat or migration, it is generally agreed that the Winnebagos dwelt in Wisconsin long before the coming of the whites and according to some investigators, they were the authors of the thousands of mounds, many of which are effigy in form, scattered over the surface of southern Wisconsin.[53] The great number of these mounds and the age of the trees growing upon many of them indicate occupancy by their builders for several centuries, during which period their numbers must have been far greater than that of the historic Winnebago at the time of Nicollet's visit.

On the other hand, the theory is advanced that the ancient miners in Lake Superior were ancestors of the Algonquian tribes and that their original home might have been northwest of the Great Lakes, from which they migrated east and south. It is a well-known fact that the Chippewa, a great branch of the Algonquian family did not have control of the mining district of Lake Superior when white man first visited that country. The Chippewa met at the Soo by white men, were called by the French, "Saulteurs", "the people of the falls" and are known to have come from the east.[54] More than that, the Chippewa knew nothing of mining and had no tradition in regard to it. They, like the Indians of all tribes, admired the shining nuggets and treasured them as divinities.

In 1825, Governor Cass was ordered by the Federal Government to assemble the Sioux, Chippewa, Winnebagos, Menominees, Sauks and Foxes at Prairie du Chien to fix and settle upon the boundary lines between respective tribes. There was but little trouble in doing this, except between the Sioux and Chippewa. The Sioux

[52]Thomas, Cyrus, Some Suggestions in regard to Primary Indian Migrations in North America, Fifteenth International Congress of Americanists, Proc. pp. 189-205; Handbook, Vol. 2, p. 578.
Curtis, Edward S., The North American Indians, 1908, Vol. 3, p. 4; and Vol. 4, pp. 129-130, attempts to combine the two theories by assuming an eastern and southern migration of certain Siouan branches north of Lake Superior before the occupation of Wisconsin and the upper waters of the Mississippi and the Missouri.
Wis. Hist. Coll's., Vol. 3, p. 285; Vol. 13, pp. 458 and 466. Jour. American Folk Lore, Vol. 26, p. 300.
Kellogg, Louise P., The French Regime in Wisconsin and the Northwest, p. 73.

[53]West, Geo. A., Indian Authorship of Wisconsin Antiquities, Wis. Arch., Vol. 6, No. 4, pp. 244-252.

[54]Davidson, John Nelson, Missions of Chequamegon Bay, Wis. Hist. Coll's., Vol. 12, p. 435.

claimed the country to Lake Superior and along it at least as far as Keweewenon Point; while the Chippewa claimed it as far south from that lake as to the St. Peter's or Minnesota, and Chippewa rivers. When the Sioux were asked upon what ground they claimed the country in dispute, they answered, "by possession and occupation from our forefathers,"—as the whites would say, "from time immemorial".

This was literally true, so far as historical reports go, for some two hundred and sixty years ago, the Sioux pursued and attacked their foes as far east as Sault St. Marie.

The Chippewa, when asked the same question, replied: "We claim it by conquest. We drove them from the country by force of arms and have since occupied it".[55]

At the time of the visit by Nicollet, probably 1639, the Menominis, a branch of the Algonquian family, were enjoying a peaceable existence at the mouth of the Menominee River, which flows into Green Bay, Wisconsin. The time of their arrival is unknown.[56]

In the treaty with the United States in 1831, the Menominees claimed as their possession, "all the lands east of Fox River, Green Bay and Lake Winnebago, and from Fond du Lac south-easterly to the source of the Milwaukee River and down the same to its mouth". They also "claimed westerly and northwesterly, everything west of Green Bay from Shoskonabie (Escanaba) River to the upper forks of the Menominee, thence to Plover Portage of the Wisconsin, and thence up that river to Soft Maple River; west to Plume River of the Chippewa, thence down the Chippewa thirty miles, thence easterly to the fork of the Monoy or Lemonweir River, and down that river to its mouth; thence to the Wisconsin Portage, thence down the Fox to Lake Winnebago".[57]

The territory claimed by the Menominees and used by them as their hunting grounds, happens to be the district of Wisconsin where the greatest number of copper artifacts and workshops have been found.

Another interesting fact is that the Milwaukee Public Museum, during the past seven years, has excavated a number of effigy mounds showing them to be, as a rule, burial mounds. Although intrusive burials are frequently encountered, the original interment

[55]Wis. Hist. Coll's, Vol. 5, p. 391.
[56]Shea, John Gilmary, Indian Tribes of Wisconsin, Wis. Hist. Coll's., Vol. 3, p. 134.
[57]Wis. Hist. Coll's., Vol 2, p. 415.

is usually found. These mounds and many of the original conical tumuli, which have been explored, produce a limited amount of potsherds and broken pots that Mr. W. C. McKern, Associate Curator of Anthropology at the Museum, believes to be either the products of an Algonquian culture, or to show a decided Algonquian influence. In case the pottery, thus recovered, as a result of later explorations proves to be Algonquian, it may re-open the question as to the authorship of the mounds.

From the great numbers of stone implements found in Wisconsin, it seems certain either that this territory has been occupied by successive tribes of Indians for a long period of time or that the ancient population must have been far greater than was ever known to white man.

Therefore, until further investigations are made and these problems solved, it would seem that the identity of the miners and of the fabricators of copper artifacts cannot be definitely determined.

Part III

ABORIGINAL COPPER ARTIFACTS

THEIR TYPES AND USES

Large numbers of native copper artifacts from the Lake Superior Region are preserved in the various museums of Wisconsin as well as in other museums of prominence throughout the country.

Among the larger Wisconsin museums, containing fine collections of native copper artifacts from the Lake Superior Region, are the Milwaukee Public Museum, possessing 1,490 implements and 1,168 ornaments, including beads; the State Historical Society Museum at Madison, with 2,000 implements and 300 ornaments, and the Logan Museum at Beloit, having 376 implements, 165 ornaments, and 5 strings of beads. The exhibits of smaller museums and private cabinets, as well as a number of single pieces, known to be still in the hands of the original finders, are none the less interesting.

Outside of Wisconsin, the collections from this state include that of the Smithsonian Institution at Washington, D. C., which is reported as containing 189 implements, 3 crescent-shaped objects, 2 lots of beads and several other ornaments. The Field Museum of Natural History at Chicago owns 584 copper implements and ornaments. The Museum of the University of Pennsylvania possesses 560 articles. The Museum of the Minnesota Historical Society has 332 implements from Wisconsin, 38 from Minnesota and 29 from Michigan, beside a number of ornaments which were secured from these states in about the same proportion. In fact, copper artifacts from Wisconsin are to be found in nearly every important museum east of the Rocky Mountains.

Mr. P. O. Fryklund of Roseau, Minnesota, has furnished the writer with drawings and data of twenty-four copper objects from eight counties of northern Minnesota, a country that may produce many more when under a greater state of cultivation. Collecting in Wisconsin has been followed vigorously for more than fifty years, yet the supply is far from being exhausted.

The awakening of an interest in archeology and the cultivation of new lands did, for a number of years, result in an annual increase in the number of "coppers" recovered. In 1876, Prof. Jas. D. Butler gave the number of Wisconsin copper implements owned by the State Historical Society at Madison as 109, the Smithsonian Institution, 30, Wisconsin Natural History Association, 14, Dr. Increase A. Lapham, 11, Milton College, 4, and Beloit College, 1.[58]

Up to 1881, Mr. F. S. Perkins of Burlington, Wisconsin, had personally collected nearly 400 copper implements. In 1886 he had increased the number to 600 specimens, indicating the rapid advance made in collecting these artifacts.[59]

The native copper artifacts of Wisconsin are properly separated into two general classes, designated as implements and decorative objects.[60]

Implements are divided into groups, each of which has been given a name suggestive of its probable use. The principal groups are designated as spear and arrow-points, knives, celts, piercing implements, crescent-shaped objects, fishing implements and spatulas. Each group is again subdivided into distinctive types.

Opinions differ among archeologists as to the probable use of many of these artifacts. The fact, however, remains that the purposes for which these objects were employed are of secondary importance as compared with the skill and ingenuity displayed in their fabrication.

SPEAR AND ARROW-POINTS

Of all the copper artifacts of utility recovered in the Lake Superior District, those commonly referred to as spear and arrow-points are the most common. They vary materially in size and shape, yet usually represent well-defined types. While they are classified, according to their size, as spear and arrow-points, it is not always possible to determine, with any degree of certainty, for which purpose some of them were intended. A few may have been supplied with short handles and used as knives. It seems unlikely that the Indian would have pointed his arrows with copper to any great extent, because of the danger of loss in shooting them, and especially when he was able, with little effort, to fabricate points of stone which would answer the purpose fully as well.

[58]Address delivered by Prof. Jas. D. Butler in 1876 before the Wisconsin Historical Society, Wis. Hist. Coll's., Vol. 7, p. 85.
[59]Hoy, Dr. P. R., Who Made the Ancient Copper Implements?, p. 13.
[60]NOTE: Unless otherwise indicated, all objects hereinafter described are Wisconsin finds and measurements of their widths refer to their widest part.

Because of their shape and characteristics, these weapons may well be classed as eyed tang, spatulate tang, notched tang, toothed tang, socketed tang, conical points, long rat-tailed tang, and short rat-tailed tang.

Eyed Tang
Plate XX

This form is rare in Wisconsin and is so named from the notch at the extremity of its tang, which suggests an open eye. The tang is flat and tapers from the blade to a narrow, square end, containing the eye.

Figure 1 shows a point from Winnebago County, four and one-half inches in length, one inch in width with a tapering base and the eye quite pronounced. The blade is flat with a rounded point.

Figure 2 shows a point from Winnebago County, five and one-half inches long, one inch wide with a flat, thin blade and tapering tang containing the characteristic eye or notch.

Figure 3 shows a specimen from Outagamie County, seven inches in length, one and one-half inches in width, with a flat, thin blade, and tapering tang, containing a depression at its extremity.

Figure 4 shows a point from Winnebago County, seven and one-half inches long, one and one-fourth inches wide, with the usual tapering tang, which is one-fourth of an inch in thickness, and shows a depression at its extremity. The blade gradually tapers to a thin cutting edge and point.

A point of this type, of native silver, two and three-fourths inches long and three-fourths of an inch wide, collected by Mr. S. D. Mitchell, from Green Lake County, is in the Museum of the University of Pennsylvania.

Spatulate Tang
Plate XX

This form differs from the last described in having a flat tang, expanded at its extremity and with a rounded end. Some are provided with a median ridge, which traverses one side of the blade from tang to point.

Figure 5 shows a specimen from Shawano County. It is eight and one-fourth inches in length, one and three-eighth inches in width, with the end of its tang rounded and considerably expanded. It has the added feature of a median ridge, which traverses one side of the blade from the base of the tang to the point of the blade. The center of the spear is one-fourth of an inch thick and bevelled on either side to a sharp cutting edge.

Figure 6 shows a specimen from Sheboygan County. It is seven and three-fourths inches long. Its blade, one and three-fourths inches in width at the widest part, has exceedingly thin edges. The tang is three inches in length with a rounded end.

Figure 7 illustrates a point from Sheboygan County. It is five inches in length, one and three-fourths inches in width, with a short tang and a convex end.

Figure 8 shows the same type as the preceding specimens and is three inches in length, one and one-fourth inches in width, with the usual spatulate tang.

Figure 9 shows a specimen from Shawano County. It is three and one-fourth inches long, one inch wide, with a slightly expanded tang and straight, flat shank.

Figure 10 illustrates a point from Fond du Lac County which seems to be a modification of this type. It is two and three-fourths inches long, one-half inch wide, with a flat tang narrowing to a slightly rounded point.

Figure 15 shows a point from Fond du Lac County. It is five and one-half inches long and one and one-fourth inches wide. The tang has a base but slightly expanded. The entire specimen is considerably eroded.

An example in the Joseph Ringeisen, Jr., collection, Milwaukee, has a perforation through the broadened part of the tang, seemingly too large for the reception of a rivet.

Notched Tang
Plate XX

This form contains a notch on each side of the base near the blade, much the same as is usually found in stone arrow-points.

Figure 11 shows a fine example of this type, two and one-half inches in length, and three-fourths of an inch in width.

Figure 12 shows a specimen two and one-half inches long, three-fourths of an inch wide, and crudely made.

Toothed Tang
Plate XX

This form of implement is quite frequently encountered in Wisconsin. It is distinguished by being generally flat or nearly of uniform thickness throughout, with a tang slightly tapering towards its extremity, the end of which is either square or rounded. The sides are provided with opposite pairs of notches or indentures, varying in number from one to six. It has been suggested that the object of these notches was to make more secure an attachment to the shaft or handle by wrapping with sinew or thongs of rawhide. This would seem conclusive, were it not for the fact that the notches, as they near the end of the shank, are closer together, leaving only those spaced most widely to serve for receiving the wrapping. The shape of these implements would suggest their possible use as knives.

Figure 13 shows a specimen two inches in length, three-fourths of an inch in width, and one-eighth of an inch in thickness at its center. Its blade has three indentures on one side and two on the other. Like many of these specimens, a median ridge traverses either side from base to tip.

Figure 14 illustrates a specimen from Shawano County. It is three and five-eighth inches in length, three-fourths of an inch in width, one-eighth of an inch in thickness and has a median ridge on either side. The tang is a continuation of the blade, separated from it by two notches, one on either side, and narrowing to a rounded point. The shape is unusual, but is not found in sufficient numbers to entitle it to a class by itself.

Figure 16 shows a specimen from Washington County. It is five and three-fourths inches long, and one and one-eighth inches wide, with a sharp, flat base, having three notches on either side. It is traversed by the usual median ridge, giving it a thickness, at the middle of three-sixteenths of an inch.

Figure 17 shows a specimen from Outagamie County. It is five and three-quarters inches long, and one and one-quarter inches wide, with thin edges terminating in an elongated shank that is supplied with a notch on either side. This form is sometimes classified as barbed or pronged-pointed.

Figure 18 shows a leaf-shaped specimen, four and one-half inches long, one and one-half inches wide and three-sixteenths of an inch thick at its middle. It has four notches on each side of its tang. A perforation or rivet hole passes through the implement where the tang and blade meet.

Figure 19 shows a specimen from Shawano County. It is three and five-eighths inches long, one inch wide, and three-sixteenths of an inch thick at its middle. It has a median ridge and the tang bears three pairs of notches. A fine series of this type from Wisconsin, is in the collection of Mr. Phillip C. Schupp, Jr., of Ravenswood, Illinois.

Socketed Tang
Plate XXI

These points are so named because the edges of the metal of the tang are turned up and over on either side, forming a substantial seating for a handle or shaft. The blades are usually strongly made and often of considerable size and weight. Some are provided with a well-defined median ridge, extending along one side, usually from end to end. While this gives added strength, it more likely was intended as an ornamentation. Some contain rivet holes near the end of the tang. A few have short bevelled edges, and an occasional one is ornamented with incised lines.

Figure 1 shows a spear of this type from Shawano County, nine inches long and one and one-fourth inches wide, at which point it is three-sixteenths of an inch in thickness. An inch and a quarter from the end of the tang the metal is one-half inch thick. A median ridge on the upper side extends from end to end, the underside being flattened. The inner side of the shank is somewhat offset, forming a shoulder against which the shaft could solidly rest.

Figure 2 shows a specimen from Fond du Lac County. It is six and three-fourths inches long with a socketed shank and without a median ridge. Its width is one inch and it has the usual offset. One inch from the end of the tang the metal is one-half inch thick.

Figure 3 shows a specimen from Shawano County. It is four and one-half inches long, and one inch wide, with a rather broad, socketed, square-ended tang, containing a rivet hole. It has the usual offset for the support of a shaft. One inch from the end of the tang, the metal is three-eighths of an inch in thickness.

Figure 4 shows a specimen from Waupaca County, three and

three-fourths inches in length and one and one-fourth inches in width, with a leaf-shaped blade, and a socketed tang, having the usual offset. It has a square end and contains a rivet hole. Half an inch from the end of the tang, the metal is one-fourth of an inch in thickness.

Figure 5 illustrates a specimen from Winnebago County. It is seven inches in length and one and one-half inches in width, having a socketed tang, half an inch from the end of which the metal is one-fourth of an inch in thickness. The end is blunted, probably from erosion. This object is peculiar in having a flat ridge extending from the socket to the point, averaging one-half of an inch in width, from which a bevel on either side extends to the cutting edge.

Figure 6 shows a specimen from Washington County. It is seven and three-fourths inches in length and one and one-fourth inches in width, with a broad shank and the usual offset. At the end of the tang is an indenture, which probably was part of a rivet hole that it once possessed. This spear-point is strongly made and capable of receiving a large shaft. The metal of the tang, half an inch from its end, is five-eighths of an inch in thickness.

Figure 7 shows a specimen from Fond du Lac County. It is nine inches in length and three-fourths of an inch in width. The blade is flat, very narrow, and drawn to a sharp point ,its upper part being incised with a zig-zag design which follows down the blade for about two inches and is the emblem of lightning. The socketed tang is considerably eroded and the metal, half an inch from its end, is one-fourth of an inch in thickness.

Figure 8 shows a specimen from Sheboygan County. It is three and three-fourths inches in length and three-fourths of an inch in width. It is symmetrical in shape, with a median ridge on its upper side, extending from one extremity to the other. The metal of the socketed tang, half an inch from its end, is three-eighths of an inch in thickness.

Figure 9 shows a specimen from Shawano County. It is a type having a short blade and long socket. Its total length is three and one-fourth inches, the blade being three-fourths of an inch wide. The socket is two and one-half inches in length, and the metal, one-fourth of an inch from its end, is one-half of an inch in thickness.

Figure 10 shows a point from Sheboygan County, much like the specimen last described. It is three and one-half inches long and one-half inch wide, with a socket about as long as the blade. The metal, one-fourth inch from the end, is one-fourth inch in thickness.

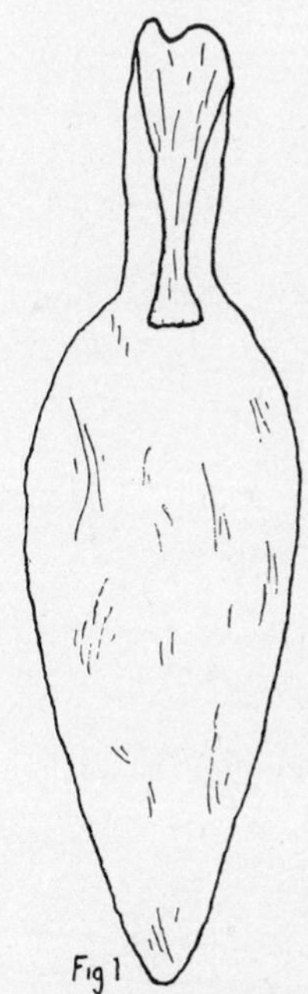

FIG. 1—Socketed copper point with leaf-shaped blade.

The smallest socketed point of this type, one-half inch in length, from Green Lake County, is in the collection of the Museum of the University of Pennsylvania.

Text figure 1 shows a form of socketed point with leaf-shaped blade, from Racine County, collected by the writer. It is six and one-half inches long, nearly two inches wide, and has a rivet hole at the end of the base.

In 1874, at Lake Five, Waukesha County, there was found a socketed point of the last described form, containing a copper rivet in place.

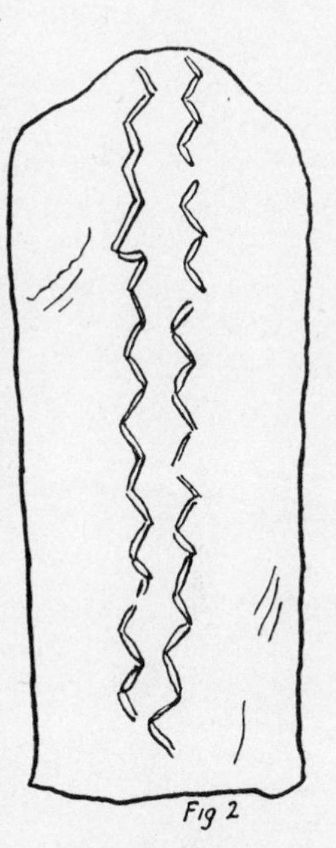

FIG. 2—Long socketed copper spud.

Examples of socketed points, with leaf-shaped blades, are to be seen in the museums at Milwaukee, Madison, and Beloit, and in almost every collection of any size from this district.

Conical Points
Plate XXI

Conical points have been collected in considerable numbers from the village sites along Lake Michigan and some of the waterways of this state, Upper Michigan, and a few from Minnesota. These points vary from one-half inch to more than six inches in length; the majority, however, are of small size. Their shape is that of an attenuated hollow cone, many of them with a solid point running an inch or more back from the tip. A few have open, angular sockets with a square or round rivet hole near the base, either for the reception of a rivet or the attachment of a line. Some may have been used as arrow-heads, others as spear-points, or even as perforators in working soft materials.

Many conical points contain a central ridge or projection, against which the handle or shaft might rest. Some are provided with two rivet holes, one above the other. The need of two rivet holes is not apparent, unless one was intended for the attachment of a line.

Figure 11 shows a specimen one and one-fourth inches long and three-fourths of an inch wide, with an open cone. The metal at the base is one-fourth of an inch in thickness.

Figure 12 shows a specimen two and one-fourth inches long and three-fourths of an inch wide with an open cone. It has a rivet hole near the base, at which place the metal is one inch in thickness.

Figure 13 shows a specimen two and one-half inches long and one-half inch wide, with a flat point. The edges of the base are turned over, forming a socket, the metal at the base being one-half inch in thickness.

"The largest example known to have been found in Wisconsin measures thirteen inches in length. It is in the E. C. Perkins Collection".[61]

In the Hamilton collection, State Historical Museum, Madison, is an unusual example of this type, seven and one-fourth inches in length, from Manitowoc County. Its blade is ornamented with a series of nine indentations.

Long Rat-tailed Tang
Plate XXI

The blade of these objects is elliptical in outline and rather flat. The tang is usually long, circular in section and tapers to a point, probably for the purpose of insertion into a handle or shaft. A few are found to have a median ridge, extending their entire length. Occasionally one has a lanceolate blade with a slightly serrated edge. They usually run in length from four to nine inches. As examples are found in nearly every collection of copper implements, they cannot be considered a rare type.

Figure 14 shows a specimen from Waupaca County. It is seven inches in length, one and one-quarter inches in width, with

[61]Brown, Chas. E., Wis. Arch., Vol. 3, No. 2, p. 78.

a rounded tang running to a point. In the State Historical Museum at Madison is an object of this type, eleven and three-quarters inches long, and one and seven-eighths inches wide.

Mr. Lee R. Whitney, of Milwaukee, is the owner of another example, five and one-half inches long and seven-eighths of an inch wide, with bevelled edges. It is unique in being about 90% native silver. This object was found at Delafield, Wisconsin. In the collection of Mr. Phillip C. Schupp, Jr., of Ravenswood, Illinois, is an example from Marinette County, eleven and one-half inches long.

Short Rat-tailed Tang
Plate XXI

This form differs from the preceding, in having a short circular tang, tapering to a point. The widest part of the blade is at the beginning of the tang.

Figure 15 shows a specimen from Fond du Lac County, eight and one-half inches long, and one and three-fourths inches wide. Its narrow, circular tang is one and one-half inches in length and one-fourth of an inch in thickness where it leaves the blade.

Figure 16, from Waupaca County, is eight and one-half inches long and one and one-fourth inches wide, with a circular tang, two inches long. The thickness of the metal, one inch from the tip of the tang, is three-sixteenths of an inch.

KNIVES

Excepting spear and arrow points, copper knives are the most common class of artifacts in this state. There are several distinct types of these cutting implements with many modifications, in fact, the Indian originated designs to suit his needs the same as the white man has done.

Copper knives can well be classified as curved back, straight back, handled, socketed, and double-edged.

Curved Back
Plate XXII

This type of knife is so classed, because of the backward curve of its blade. The tang is usually short and tapers to a point for the reception of a handle. They range in length from one to twelve inches.

Figure 1 shows a specimen from Fond du Lac County. It is an exceptionally long example of the knives found in Wisconsin, yet of a common type. The length is twelve and one-quarter inches, the width, one and one-quarter inches, and the thickness of the metal beginning at the tang, one-eighth of an inch, with its back slightly curved.

Figure 2 illustrates a specimen from Manitowoc County. It is of the same type excepting that for some reason the end of its tang is curled. The length is nine and three-quarters inches, the width, one inch. This seems to be a modification of the curved back type, as the back is but slightly curved.

An exceptionally fine example in the Sawyer Museum at Oshkosh is seventeen and one-half inches in length and weighs eleven ounces. A small example of this type from Sheboygan County is in the R. Kuehne collection.

Many knives of this form have blades ornamented with indentures. An example in the collection of Mr. Frank Miller, of Princeton, Wisconsin, is four inches long, one inch wide, and has eight deep circular indentures on one side and ten or twelve on the opposite side.

Figure 7 shows a fine example ten and one-quarter inches in length. The tang near its rounded end is five-eighths of an inch in width.

Figure 12 shows a specimen from Fond du Lac County. It has a pointed tang and a curved, almost crescent-shaped blade. It is two and three-quarters inches in length; the blade is three-quarters of an inch wide and the metal at the base of the tang is one-eighth of an inch in thickness.

Figure 13 shows a specimen of the same type, four and three-quarters inches long and one inch wide. The metal at the back of the blade is one-quarter of an inch in thickness.

Figure 14 illustrates a knife from Shawano County, which is an exceptionally large specimen of this type. It is nine inches long with a curved blade, the width of which is one and one-half inches, the metal at the base of the tang being one-eighth of an inch in thickness.

Straight Back
Plate XXII

This type of knife has a straight or nearly straight, thick back, with a flat or pointed tang, for the reception of a handle.

Figure 3 shows a modification of the type above described, having a shorter and wider blade, with a straight back. The length is eight inches, the width, one and one-half inches and the metal at the base of the tang is three-sixteenths of an inch in thickness.

Figure 4 shows a specimen from Fond du Lac County. It has a total length of six inches, a length of blade of three inches and a width of three-quarters of an inch.

Figure 5 shows a specimen from Hancock, Michigan, representing a still greater modification of this type, being four and one-eighth inches long and one inch wide.

Figure 6 shows a specimen from Shawano County. It has a broad tang, which indicates that the handle was formed by a split stick into which the tang was fitted and then wrapped with rawhide or some other material, to make it secure. This specimen is six inches in length and one and one-quarter inches in width, the tang being one-half inch wide at its rounded end.

Figure 8 shows an unfinished knife, apparently cut out of a sheet of copper, but not hammered or ground to an edge. It is another example of the same type. It is six and one-quarter inches in length and from one-eighth to three-tenths of an inch in thickness. The edges are square all around. It was found on the surface near Hancock, Michigan. In the same collection are three straight back knives, from the same location, averaging but one and one-half inches in length.

Handled
Plate XXII

This type is characterized by not requiring the addition of a handle.

Figure 11 shows a knife with blade and handle in one piece. The length is six and one-quarter inches, the width of handle, one inch, and the thickness, one-eighth of an inch. It has a straight back with a sharp point and a convex cutting edge of two and one-half inches. The handle part was probably wrapped with some material to make it better fit the hand when in use.

Almost a duplicate of the last described knife was found near Wampun, Manitoba, by Mr. Ole Pederson of that village. Another of the same type, four inches in length, was secured near Drayton, North Dakota, and is now owned by Mr. Helmer Stenquist of that place.

Figure 15 shows a knife from Washington County. It is a specimen of rare form which is ten inches in length and has a width of blade of one and one-half inches. It has a square, straight back, bevelled to a cutting edge and ornamented by six indentures on each side. About four and one-half inches from its point, the object narrows to one-half inch in width for a distance of four inches, forming a handle. Beyond this, it widens to one and one-quarter inches, and terminates in a square, cutting edge. It is also provided with a cutting edge on the right side of the base. The handle part of this specimen was doubtless wound with some material, making it more convenient for use.

A second example of this type, but slightly smaller, is in the Milwaukee Public Museum.

In the collection of Mr. J. T. Reeder, of Houghton, Michigan, is a specimen of this type, with the back slightly curved. It was secured near the Lake Superior Canal, in Houghton County. A fine example from Waupaca County is in the collection of Mr. Phillip C. Schupp, Jr., of Ravenswood, Illinois. This implement is ten and one-half inches long. The end of the handle curves upward toward the back, thus making the grip more secure.

Socketed
Plate XXII

This form of knife is provided with a socket, usually containing a rivet hole, for the reception of a handle, which in some instances may have been of considerable length.

Figure 9 shows a specimen from Trempealeau County. This knife has a socketed shank with a rivet hole. It is nine inches in length; has a greatest width of one inch and a straight back and socket for the attachment of a handle. The thickness of the metal at the socket, half an inch from its end, is three-eighths of an inch.

A similar example is in the collection of the Neville Public Museum, Green Bay.

Double-edged
Plate XXII

These objects are usually leaf-shaped, with a thin, flat blade which is bevelled to a sharp cutting edge on each side. They are in the form of a dagger and are often provided with one or two rivet holes. Many of the leaf-shaped implements, classified as spear or arrow points, were doubtless used as knives.

Figure 10 shows a specimen six inches long and one and one-half inches wide, with a double cutting edge and a square-ended tang, containing two rivet holes.

The State Historical Museum at Madison has about two hundred knives in its collection; the Milwaukee Public Museum one hundred sixteen; the Logan Museum at Beloit, forty-five; and the Minnesota Historical Society, twenty-eight from Wisconsin, two from Minnesota and two from Michigan. In fact these implements are to be seen in nearly every collection of any size.

CELTS

The name celt, applied to this class of implements, may have come from the Latin *celtis,* meaning chisel, or perhaps from being used by the Celts. Implements in stone of this general type have often been found in the barrows and other repositories of antiquarian remains in Southern Europe.

Celts of copper vary in size and shape and can be classified as spuds, axes, chisels, adzes, wedges, and gouges.

Spuds

In the lake regions of northern and eastern Wisconsin, particularly, have been found a limited number of objects, which, because of their resemblance in form to certain stone implements, are called spuds. They have a rather broad cutting surface, are socketed, and strongly made. They are from three to eight inches in length, and their blades vary in width and in outline. When in use they were provided with a strong handle, as the large sockets indicate.

Like those stone implements from which they derive their name, these tools may have been employed in stripping bark from trees. They were also doubtless used in cutting holes through the ice for winter fishing, as is indicated from the fact that one of these spuds was recovered from the deep waters of Long Lake, Iron County; it being brought to the surface, a few years since, by rod and line. A multitude of other uses could have been made of them, such as chipping charred wood in the building of dugouts, and in the clearing of land for agricultural purposes. They might well then be considered as implements of general utility.

The handle was undoubtedly driven into the socket solidly and probably made secure by wrappings of rawhide, permeated or more or less covered with pitch thus making a very firm attachment.

Spuds may be classed as long socketed and short socketed.

Long Socketed
Plate XXIII

In this form the socket is elongated and its sides reach nearly to the cutting edge, which is square or slightly convex and much narrower than the socket. In the finest examples of this type, the socket terminates in a point at one end and a median ridge traverses the back from this point nearly to the cutting edge.

Figure 2 shows a specimen five and one-half inches long in which the top of the socket is two and one-fourth inches wide. The point is square, one and one-half inches wide, and shows much use. The thickness of the roll of metal at the end of the socket is one and one-half inches.

Figure 3 shows a specimen six and three-fourths inches long in which the socket is two and one-half inches wide. The implement expands at its middle to three inches in width. This blade is provided with a nearly straight cutting edge one and one-half inches wide. The thickness of the roll of metal at the end of the socket is one and one-half inches.

Figure 4 shows a specimen from Dodge County. It is seven and one-fourth inches long and the top of the socket is two and one-fourth inches wide. The cutting edge is slightly convex and flares to a width of three and one-half inches. The socket extends far down on the blade, nearly to the cutting edge. The metal throughout is thin, indicating its use more as a gouge. The thickness of the roll of metal, one and one-half inches from the end of the socket, is one inch.

Figure 11 shows a specimen from Calumet County. It is graceful in form and five and one-fourth inches in length. The widest part of the socket is two and three-fourths inches from which it narrows to two and one-fourth inches. The socket extends nearly to the cutting edge, which is slightly rounded and flares to a width of two and five-eighths inches. The upper end of the socket curves upward for one and one-half inches to a point and its rolled sides extend downward, almost to the cutting edge. The implement is strongly made and well proportioned. The metal of the roll near its top is one inch in thickness.

An example of this form, in the Milwaukee Public Museum, shown in text figure 2, has two rows of incised zig-zag lines, extending the entire length of the back.

Short Socketed

Plate XXIII

This form has a wide, short socket extending downward to within an inch or so of the cutting edge, which is usually semi-circular or crescent-like in outline. The end of the socket, nearest the cutting edge of this type of implement, ends in a shoulder against which the end of the shaft would solidly rest. This shoulder or offset brings the face of the blade on an even plane with the edge of the shaft, thus enabling the operator to cut to a line and at right angles to any depth.

Figure 8 shows a specimen from Ludington, Michigan. It is five and one-half inches in length. The width of the socket is four inches; its length, three and one-half inches. The cutting edge is convex and three and one-half inches in width. The lower end of the socket operates as a shoulder, upon which the handle might rest. The back of this implement is slightly convex and its face flat. The thickness of the roll of metal of the socket, for nearly its entire length, is one inch.

FIG. 3—Unusual form of copper spud.

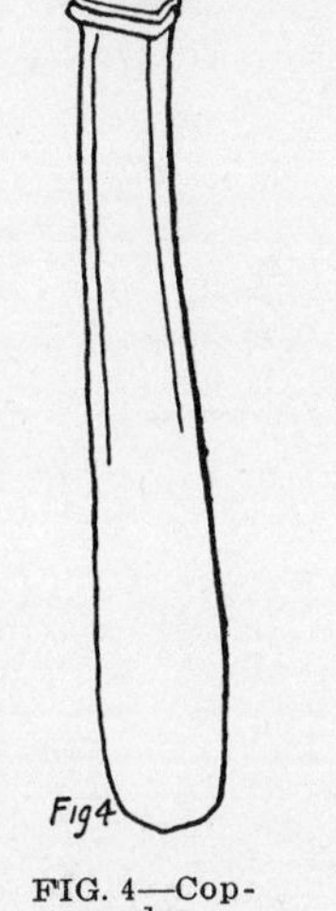

FIG. 4—Copper adze.

Figure 9 shows a specimen from Waupaca County. It is four and one-half inches long and the width of the socket three and one-fourth inches. The width of the crescent-shaped cutting edge is two and three-fourths inches. The thickness of the roll of metal of the socket is one inch. The figure is a back view of this type and distinctly shows the shoulder.

Figure 10 shows a specimen from Iron County. It is four and one-half inches in length, its base being three and one-half inches wide. The thickness of the metal at the top of the socket is one and one-fourth inches. The convex cutting edge flares to three and seven-eighths inches in width while the blade is less than one inch in length and the shoulder absent.

An unusual form of spud, shown in text figure 3 was plowed up on the top of Diamond Bluff, in the northwestern part of Wisconsin. It has a total length of fifteen and one-half inches and a width at mid-section of one and one-eighth inches although the cutting edge is less than one inch wide. The socket of this object is three inches long, circular in form and of such a width that the transverse diameter of the instrument is not disturbed.[62]

One or more of the types, shown in plate XXIII, with various modifications, are to be seen in all larger collections of copper implements secured in this district.

The Milwaukee Public Museum is the owner of thirty of these implements, the State Historical Museum at Madison has twenty-five, the Logan Museum at Beloit, fifteen, the Smithsonian Institution, thirteen, and the Minnesota Historical Society, four. Examples are to be found in the collections of other museums and private cabinets.

Axes

Not only the village sites, but the mounds of Wisconsin, have produced a large number of these artifacts. They vary in length from two to fourteen inches and weigh from half a pound to three and a half pounds. These implements must have been supplied with handles. The most probable method of hafting was by means of a stick of the desired length which had at one end a slot sufficiently deep to receive the axe. This was afterwards secured by strips of rawhide wound above and below.

Flat Blade
Plate XXIV

This type is flat, or nearly so throughout and of uniform thickness, with parallel edges tapering toward the pole. The cutting edge is convex and slightly flaring. The length varies from three to ten inches.

Figure 9 shows a specimen from Waushara County. It is eight and one-half inches long, with a convex cutting edge, three and

[62]Winchell, N. H., The Aborigines of Minnesota, p. 503.

three-quarters inches wide, which becomes gradually narrower toward the pole. This is square in form, two and one-half inches in width and one-eighth of an inch thick. Half an inch back from the pole the thickness gradually increases to three-eighths of an inch. Four inches from its cutting edge, the thickness at the middle is one-quarter of an inch.

Figure 10 shows a specimen from Trempealeau County. It is six and three-quarters inches in length, three and one-quarter inches in width and tapers toward the top which is square, two inches wide and one-quarter of an inch in thickness. At its middle, the implement is three-eighths of an inch thick. The cutting edge is slightly convex.

Bell-shaped

Plate XXIV

But few examples of this form have been recovered. For the lack of a more appropriate name, this type is usually referred to as bell-shaped. The specimens are generally flat and thin. The edges curve equally inward from a broad cutting edge to the center and then outward to an almost equally broad pole. The head is square and the cutting edge nearly semicircular.

Figure 8 shows a specimen from Milwaukee County. It is eight inches long, with a convex cutting edge, four and one-half inches in width. The pole is three and one-half inches wide. The edges curve inward and then outward from end to end, leaving a median width of two and one-half inches and at this point there is a thickness of one-fourth of an inch with a gradual taper each way which leaves the head or pole but one-eighth of an inch in thickness.

A second example of this type, an inch shorter, is in the same collection.

Grooved and Notched

Plate XXVIII

While the copper axes, recovered in Wisconsin, are not fashioned with a groove, the Milwaukee Public Museum is the owner of a grooved axe of Lake Superior copper taken from a mound on the banks of the Black Snake River, Utah.

Figure 2 shows this unique specimen. It is three and seven-eighths inches long and two and one-half inches wide, with a maximum thickness of one and one-quarter inches. It is bevelled to a sharp square cutting edge at one end and encircled with a groove near its flattened pole. This axe is evidently copied after an ordinary type of stone axe.

Mr. M. C. Long, of Kansas City, is reported to own a grooved copper axe. In the H. P. Hamilton Collection, State Historical Museum, Madison, "is a notched copper axe, which came from the vicinity of Horicon."[63]

The so-called notched axes are usually rude specimens, irregular in shape or oval in outline. They are broadest at the pole and slightly notched on each side of the margin, apparently for the purpose of giving stability to the handle by binding. These objects vary from three to six inches in length and are usually of considerable thickness. A number of examples are in the Milwaukee Public Museum. One specimen, from Dodge County, in the Hamilton collection of the State Historical Museum at Madison, is said to weigh five pounds.

Double Bitted
Plate XXVIII

So-called double bitted axes are extremely rare and were probably used as weapons. But one example of this type is known to the writer. It is in the Milwaukee Public Museum.

Figure 3 shows this specimen from Waupaca County. It is six and one-half inches in length, and has a median width of three-fourths of an inch. The central thickness is one-half inch, and it is bevelled on each end to a flaring, convex cutting edge, one inch in width. The edges curve inward and then outward from the middle portion of the object to the bits. This unusual specimen, if hafted, could have been used as a tomahawk.

In every copper collection of considerable size are to be found one or more implements which cannot be classified with any degree of certainty. Some objects, because of their shape, are thought to have been intended as clubs or bludgeons. Long, thin, curved specimens, occasionally found, suggest swords. Others, finished and unfinished, are encountered, whose exact functions are unknown.

Figure 1 (Plate XXVIII) shows an implement, from Fond du Lac County. It is six inches in length and three-fourths of an inch

---

[63]Brown, Charles E., in Moorehead, Warren K., The Stone Age of North America, p. 182.

in width, with a maximum thickness at the middle of five-eighths of an inch.

This unique specimen is bevelled to a sharp convex cutting edge at one end and to a point at the other. While it might have been used for various purposes, if hafted, it would have made a formidable weapon when used as a tomahawk.

Chisels

Among the aboriginal copper implements of utility, is a class known as chisels. A large number are represented in the various collections of this district. They range in length from four to fifteen inches and in weight from a few ounces to nearly four pounds. These splendidly formed implements were doubtless employed in the excavating of canoes and the working of wood generally. Some may have been supplied with a handle and used as adzes.

They may be divided into three distinct types, known as: broad cutting edge, uniform width and median ridge.

Broad Cutting Edge
Plate XXIV

This class is distinguished by having a broadened cutting edge, from which the implement gradually narrows to a square or rounded edge. The blade is usually thickest at the middle, gradually tapering toward its extremities. The upper surface is often convex and the lower flat. The edges may be either rounded or flat.

Figure 1 shows a specimen from Milwaukee County. It is a fine example, ten inches long, with a flaring cutting edge, three and one-half inches in width. The smaller extremity is square, one and one-quarter inches wide, and three-eighths of an inch thick. At its middle, it is one-half of an inch thick.

Figure 2 shows another specimen of the same type, nine inches in length with a cutting edge, two and one-quarter inches wide. It is one-half inch in thickness at the center, and the smaller extremity is turned up, as if intended for scraping purposes. It gradually narrows from the cutting edge to the other extremity.

Figure 3 shows a specimen from Calumet County. It is eight and three-quarters inches in length, with a cutting edge two inches wide. The width of the implement gradually increases toward the smaller end for a distance of two inches, at which point the center is three-eighths of an inch in thickness. From this point it tapers to the smaller end, which is five-eighths of an inch in width and three-sixteenths of an inch in thickness. This has all the appearance of being a large flesher.

Figure 7 illustrates a specimen from Ozaukee County, five and one-eighth inches long and two and one-half inches wide at its cutting edge, which is nearly square. From this edge the sides gradually taper to the other extremity. The smaller end is slightly battered, three-quarters of an inch wide, and one-quarter of an inch in thickness.

In the Neville Public Museum at Green Bay is a fine example, from Shawano County, eleven and three-quarters inches long and nine-sixteenths of an inch thick. Its convex cutting edge is two inches wide.

Uniform Width
Plate XXIV

This type is either of uniform width throughout its length, or narrows slightly from the cutting edge to the smaller end. It is characterized by having a cutting edge at each end. This class of implements may have been used as adzes, or, in the hand, without the addition of a handle.

Figure 4 shows a specimen eight and three-fourths inches long, one and one-fourth inches wide at its greater cutting edge and three-fourths of an inch wide at the opposite extremity. It is one-half inch in thickness at the middle, tapering either way to a cutting edge.

Figure 6 shows a specimen from Waupaca County, which is an unusual example of this type. It is thirteen and one-fourth inches in length, one and three-fourths inches in width at the larger end, and three-fourths of an inch in width at the smaller extremity. For some distance along the middle, it is one-half inch in thickness.

Median Ridge
Plate XXIV

This is a rather uncommon type that has nearly straight, parallel sides with a slightly flaring cutting edge. Its extremities are rounded and the upper surface is traversed by a well-defined median ridge which extends to within a short distance of the cutting edge and makes a very strong implement.

Figure 5 shows a specimen from Calumet County. It is a fine example of this type, nine and one-half inches in length, two and one-eighth inches in width at the cutting edge and one and one-half inches wide at the other extremity. At the beginning of the cutting edge, the chisel is three-quarters of an inch in thickness and at the smaller end one-quarter of an inch in thickness. There is no reason to believe that this tool had a handle when in use.

ADZES

These implements might be classed as chisels, were it not for their being slightly curved from end to end. Hence, when supplied with handles, they could be used without any portion of the handle striking the surface upon which work was being done. These objects either have a square or slightly convex cutting edge and gradually narrow toward the other extremity, which prevents the handle from working toward the cutting edge when in use. The bevel of the cutting edge is entirely on the inner side, the same as in adzes used by white men.

Text figure 4 shows a specimen in the Milwaukee Public Museum which is seven and one-third inches long, with a cutting edge, one inch wide.

A slightly shorter example of the last described form is to be seen in the same collection.

Drawings and data of three most interesting celt-like objects of native copper were furnished by Mr. Phillip C. Schupp, Jr., of Ravenswood, Illinois. They were part of a cache of thirteen objects, some of which are beveled to a cutting edge on one side only and are doubtless adzes. This cache was found in Oneida County, Wisconsin, and the specimens are now in the collection of Mr. Schupp.

The largest specimen is eighteen and one-quarter inches long, two and one-half inches wide and one-half inch thick at its center, which thickness continues for a considerable part of the length. The sides are square and gradually taper to a slightly convex cutting edge. The face is flat. The back is also flat to within four inches of the cutting edge, which is formed by a taper or bevel. The cutting edge is nearly on a plane with the face, but is slightly concave, much like an adze. The opposite extremity is square, one inch wide and a quarter of an inch thick. The entire adze weighs five and three-eighths pounds.

Another implement is of the same form and weight but is one inch shorter than the one above described.

The third example, of the same type, is thirteen and one-half inches long and weighs three and one-quarter pounds.

These artifacts are difficult to classify, as they may have been used as adzes, chisels, picks or agricultural implements.

In the same collection, is the blade of an adze, five and one-half inches long. The cutting edge is one and threee-quarters inches broad and beveled from one side only, as is the case with most copper adzes. From the cutting edge the object tapers to a width of three-quarters of an inch at the opposite end. This is broadened at the extremity, and on one side ends in a sharp point one-half inch in length and at right angles to the blade.

A similar example is in the collection of Mr. Vetal Winn, of Milwaukee.

This modification of the type is rarely encountered, and was probably intended to give stability to the handle.

WEDGES
Plate XXIII

In nearly every archeological collection of any size throughout the Lake Superior District, are to be found copper implements which may properly be termed wedges. They are usually from three to five inches in length and taper from about one-half inch in thickness at one end, which is square, to a rather sharp edge at the other extremity. The width is generally from an inch to an inch and a half, and often the same for the entire length of the implement. The cutting edge is usually straight but sometimes is widened. The battered square ends of many of these implements show much use. If wooden mauls were used in driving these wedges, the ends would not be thus disfigured, which may account for the perfect condition of many of them. It is reasonable to suppose that these wedges were used for the purpose of splitting wood and, probably, to some extent, in mining.

Figure 5 shows a specimen from Waupaca County three and three-fourths inches long. The thickness, one-half inch at the narrowest part, tapers toward the rounded cutting edge which is one and three-fourths inches in width.

Figure 6 shows a specimen from Marquette County, four inches long and one and one-fourth inches wide throughout its entire length. The thickness is one-fourth of an inch near the square end and tapers to a straight cutting edge.

Figure 7 shows a specimen from Hancock, Michigan, three and one-half inches in length. It has a battered, square top, one and one-fourth inches wide and one-half inch thick. It tapers to a convex cutting edge, about one inch in width.

GOUGES

These objects are distinguished from chisels by a concavity which extends a considerable distance back from the cutting edge and occasionally occupies the entire length of the implement. Some examples are of the same width throughout, with a straight cutting edge, while in others the width gradually increases toward the cutting edge, which is flaring and slightly convex. The inner face of the object is sometimes bordered on either side by a ridge, which curves inward and may have been intended for the reception of a straight shaft as a handle. The metal is usually thin, indicating its use as a wood-working tool.

The two types may be classed as expanded cutting edge and uniform width.

Expanded Cutting Edge
Plate XXIII

Figure 4 shows a specimen from Dodge County, seven and one-fourth inches long. The top of the socket is two and one-fourth inches wide and one inch thick. The cutting edge is nearly square and flares slightly to a width of three and one-half inches. The socket extends nearly to the cutting edge. The back of the object curves outward and the metal is thin.

Uniform Width
Plate XXIII

Figure 1 shows a specimen from Fond du Lac County, seven and one-fourth inches long, with a slightly convex cutting edge, and three and one-fourth inches wide. The top is one and three-fourths inches in width. At the center the thickness is three-fourths of an inch. The sides are nearly straight, yet slightly flaring near the cutting edge. The inside is concave, suggesting use as a gouge but it might have been used as an adze.

The distinguishing feature between an adze and a gouge might be the manner in which the handle was attached, that of a gouge being a straight shaft the same as that of a spud, to be used with both hands, while an adze was probably supplied with a short handle, nearly at right angles to the blade, and used with one hand.

An example from Calumet County, is seven and four-fifths inches long, one and one-fourth inches wide and of a uniform width throughout. The back curves outward and the front is concave. The metal of the parallel sides is turned to form a concavity extending the entire length. It has a straight cutting edge at each end. This implement was possibly used without a handle.

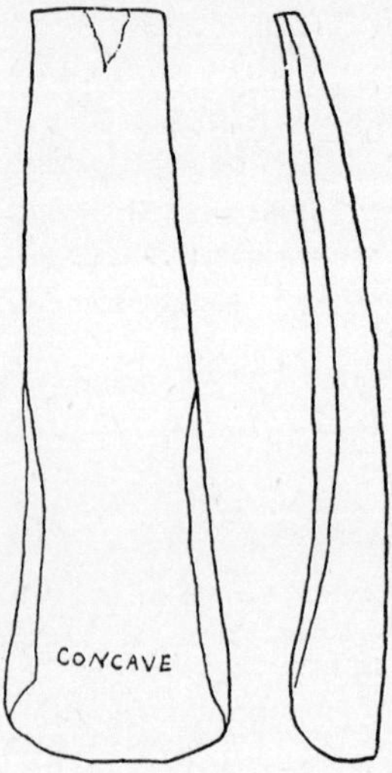

CONCAVE

FIG. 5—Copper gouge of uniform width.

Test figure 5 shows an example from Calumet County, six and one-half inches long. The cutting edge is somewhat flaring and one and seven-eighths inches wide. From it the object tapers to a straight, blunt base, one inch wide and somewhat battered from pounding. This specimen is particularly interesting because of a triangular piece of copper that has been set into the base and made a part of the whole by annealing and welding.

Fig 6

FIG. 6—Copper harpoon with long tang.

PIERCING IMPLEMENTS

This class of implements includes pikes, punches, drills, awls, and needles.

Pikes
Plate XXVI

Pikes are rod-like in form, usually circular, sometimes square and occasionally rectangular in section. They generally taper to a point at each end. In a number of examples, however, only one end is pointed, the other being square. The largest specimen which came to the writer's notice, is at the Field Museum at Chicago. This extraordinary example is nearly forty inches in length, one inch in diameter at the middle, and tapers to a point at either extremity. It weighs five and one-fourth pounds.

In 1873, Mr. Fred Perkins, of Burlington, Wisconsin, reported that he had owned a specimen of this type found near Barton, Wisconsin, that was three feet long and nearly an inch in its greatest diameter. This remarkable object was unfortunately lost.[64]

In the State Historical Museum at Madison is an example from Outagamie County twenty-nine inches long.

That implements of this class were employed as weapons seems improbable. The larger forms were doubtless used as burning irons in making wooden "dugouts", mortars and other vessels. After burning a series of holes, the splitting of the wood would become an easy task.

Figure 2 shows a specimen from Waukesha County, twelve and one-half inches long and square in section. It is three-sixteenths of an inch in thickness at the middle, and pointed at each end.

Figure 3 shows a specimen from Shawano County, eleven and one-fourth inches in length and flat in section. It is three-sixteenths of an inch in thickness at the middle and the width of the head is one inch. One end is pointed and the other doubled back.

Figure 4 shows a specimen from Waupaca County. It also has one end doubled back. It is ten and one-half inches long, flat in section, and three-sixteenths of an inch thick at the middle.

A most interesting example is in the Joseph Ringeisen, Jr., collection, at Milwaukee. It is sixteen inches long and three-fourths of an inch square for six inches each way from the mid-section. It weighs three and three-fourths pounds. At one end the object terminates in a rounded point, the bevel commencing three inches from the end. The opposite point is reduced in the same manner to a square measuring one-half inch on each side.

A copper pike was found in Carlton County, Minnesota, in 1926. It is thirty-two and one-half inches long and three-fourths of an inch in its greatest diameter, some four inches above the point. The piece tapers gracefully to the butt, where it is bent into a claw or hook. Its weight is 4 pounds. This splendid specimen has been deposited in the museum of the Minnesota Historical Society.[65]

---

[64] Foster, J. W., Prehistoric Races of the United States, p. 259, 1873.

[65] Babcock, Willoughby M., Wis. Archeologist, new series, Vol. 7, Pt. 4, p. 218.

Punches
Plate XXV

Punches are of various shapes and lengths. They have a flat top at one end and taper to nearly a point at the other. They are usually round or square in section. Many of them, by reason of their battered tops and ends, show considerable use. Some forms are referred to locally as spikes, because of their enlarged head, tapering body and sharp point.

Figure 1 shows a specimen from Calumet County, representing a large example of a punch, seventeen and one-fourth inches in length and three-fourths of an inch in diameter at the middle. It is intermediate between round and square in form, and tapers to a nearly square point at one extremity, while the other end shows the result of being hammered by some instrument.

Figure 5 shows a specimen from Waukesha County, which, for no better name, is called a punch, although it may have been used for some other purpose. It is seven and one-half inches long, three-fourths of an inch wide and square in section. It is worked to a square point at one end and the other is turned over in the form of a hook.

An example of this type, eight inches long, was collected by the writer from the Halter village site, Racine County, in 1878.

In the collection of Mr. Phillip C. Schupp, Jr., of Ravenswood, Illinois, is an example fourteen inches long, and another eleven inches in length. Both were taken from a mound in Brown County. A third specimen eleven inches long comes from the same county.

In the collection of Mr. F. S. Dayton, of New London, Wisconsin, is an interesting form of punch, two inches long. Just below the battered top, it is one-fourth of an inch square and gradually tapers to a blunt point. About the center of the object, it curves out and then down again, thus giving the workman a clear view of his work when in use.

Drills
Plate XXV

Some of these piercing implements might, from their shape, be classed as drills and reamers. Those that are square in section would answer the purpose of drills for working in wood and other soft materials. The type that is broad at one end and tapers gradually to the other, would suggest a reamer.

Figure 6 shows one form of drill from Ozaukee County. It is five and one-half inches in length and one inch in width. It has the added feature of a median ridge or elevation traversing either side of the blade from end to end, from which the blade is bevelled to a cutting edge. It is reduced to a point at one end, while the other is irregularly convex. The greatest thickness at the middle is one-fourth of an inch.

Figure 10 shows a different type of drill which is square in section and shaped at each end. It could be set in a handle. It is three and one-half inches in length and one-half inch in thickness at the middle.

Figure 11 shows another example of the same type, four and three-quarters inches long and one-quarter of an inch in thickness at the middle.

In the collection of Mr. Frank Miller, of Princeton, Wisconsin, is a specimen of the conical-shaped type, three and one-half inches long, found near that place. The width of the socket is one inch, from which it tapers to a point at the other extremity. For one and one-half inches from the point, the object is square in section and of solid metal. This object is strongly made and was doubtless used as a drill or reamer.

Awls
Plate XXV

Awls have been obtained in great numbers from the village sites along the Wisconsin shore of Lake Michigan, as well as from those on the banks of the Fox and Wolf rivers. They vary from one to eight inches in length and from one-sixteenth to one-half of an inch in thickness. The smaller ones predominate.

These perforators are usually slender and pointed at one or both ends. They are either round or square in section, with a short or long taper to the extremities. The lighter forms are sometimes provided with a handle, as an example in the Milwaukee Public Museum, with a ferrule attached, would indicate.

A second example is in the S. D. Mitchell collection, now owned by the University of Pennsylvania. This specimen was found in Green Lake County, Wisconsin, and is set in a polished bone handle. The total length of the awl and handle is two and one-half inches.

Another example, set in a bone handle, with a total length over all of four inches was found on Rhea's Island, Loudon County, Tennessee.[66]

Copper awls and other forms of perforators, supplied with a handle when in use, are nearly always surface finds. Here decomposition is rapid, which accounts for the almost total absence of this appendage when found.

Figure 7 shows a specimen from Washington County, three and one-fourth inches long and three-sixteenths of an inch thick at the middle. It is flat in section reduced to a point at one end and broadened to one inch in width at the other, thus providing a handle that would permit the instrument to be revolved by the fingers when in operation.

Figure 9 shows a specimen from Winnebago County, three and one-fourth inches in length. It is round with one end flattened, and tapers to a sharp point at the other extremity.

Figure 8 shows a specimen two and three-fourth inches in length, round in form and nearly one-fourth of an inch in thickness at the middle. It tapers to a sharp point at one end and to a blunted point at the other. When in use it was doubtless provided with a handle.

Figure 12 shows a common form of awl, simply a very thin rounded rod. The one, shown in this figure, is two and three-eighths inches long and tapers to a sharp point at either end. The shape would indicate that a handle of wood or bone was supplied.

Figure 13 shows a different type of awl. It is a thin rounded rod, three and three-fourths inches in length. It tapers to a sharp point at one end, while the other end, though sharp, is flattened. It seems to be a perforator of some kind, and may have been supplied with a handle.

Needles
Plate XXV

Copper needles range in length from two to more than eight inches, the majority of them being about three inches long. They are provided with an eye, similar to our modern needle. In rare cases a needle is found containing two eyes, thus enabling the seamstress to use two colors of thread at the same time.

Figure 14 shows a specimen from Columbia County, five and

[66]Thruston, Gates P., Antiquities of Tennessee, 2nd ed., p. 302.

one-eighth inches in length. It is provided with one eye about one inch from the larger end, and a second one an inch nearer the middle.

In the Milwaukee Public Museum is a series of needles from Mexico of practically the same form as those found in Wisconsin.

In the Joseph Ringeisen, Jr., collection, at Milwaukee, is a fine example, four and one-quarter inches long. It has a large eye at one end, from which the needle tapers to a sharp point.

In the J. T. Reeder collection at Houghton, Michigan, is an example containing a perfect eye near one of its extremities.

The use of this implement is too well known to warrant a detailed explanation.

FISHING IMPLEMENTS

Harpoons

Copper harpoon points are of very infrequent occurrence in Wisconsin, yet a number of distinct forms have been recovered from the old Indian village sites of this state. The necessity of the harpoon here is not apparent, unless it was used for the capturing of large fish, such as muskalonge or sturgeon, which still inhabit many of the lakes and streams of the district.

These implements vary in form and can be divided into four principal types: multiple-barbed, socketed tang, long tang and short tang.

Multiple-barbed

Plate XXVI

This type of harpoon is strongly made, with a flat or rounded blade. One edge is quite thick and the opposite one from which project a number of barbs much thinner. The tang and point are each short and taper to a blunt extremity.

Figure 1 shows an implement eight and three-eighths inches in length and three-fourths of an inch in width. The back is three-eighths of an inch in thickness at the barb nearest the shank. The opposite side is quite thin and is furnished with four stout barbs set about one and one-half inches apart. The ends taper to a blunt point. This specimen is unique, as no other of its type, so far as known, has been thus far recovered in this state.

Socketed Tang
Plate XXVI

This type seldom exceeds five and one-half inches in length. The socketed tang usually contains a round or square rivet hole, probably intended for the attachment of a cord similar to the toggle type of harpoon used by the Eskimo. The blade is generally flat and tapers from the socket to a point. Usually the blade is provided with a single barb set a short distance from the beginning of the socket. Other examples have a projecting barb near the point. A few have been recovered in which the barb is produced by an extension of the metal upward from the opening of the socket.

Figure 2 shows a harpoon from Winnebago County, five and three-fourths inches long and about one inch wide at the barb. A barb projects from the edge of the blade, one inch from the socket. The socket is provided with a rivet hole, possibly indicating that a copper rivet was passed through the spear shaft in order that the harpoon might not be lost.

Figure 3 shows a specimen from Milwaukee County, five inches in length, with a base, one inch in width and three-eighths of an inch in thickness. It is of the same type as the one last described, excepting that the rivet hole is square instead of round, and the barb starts one-half inch below the socket.

Long Tang

This form is cylindrical in section and much resembles in shape the iron trade harpoons, furnished the Indians by white men. They are of large size, sometimes ten or more inches in length. The single barb is located an inch or two from the point, from which the tang gradually tapers to a sharp extremity.

A long, narrow example, cylindrical in section, and tapering to a point at each end was formerly in the W. H. Ellsworth collection. This specimen was ten and three-fourths inches in length and about one-half inch in diameter at its middle. It was provided with a pronounced barb, located about two inches from the spear point.

"Two examples of this type of harpoon or fish spear point are in the collections of the State Historical Museum, the larger of these being nine and seven-eighths inches long. The base of the pointed half-inch barb is two and one-half inches from the sharp tip of the implement. This point was found in Merton Township, Waukesha County, in 1877. It weighs four ounces.

"The other specimen is eight and five-sixteenths inches long and weighs three ounces. The barb is within seven-eighths of an inch of the tip of the implement. It is very short, being one one-eighth of an inch long. This point was found at Lake De Neveu, Fond du Lac County."[67]

The Logan Museum, at Beloit, is the owner of two similar specimens, one eight and three-quarters inches long and the other an inch shorter. The end of the tang of one of them is bent in the form of a small hook as shown in text figure 6.

In the collection of Mr. J. P. Schumacher, Neville Public Museum, at Green Bay, Wisconsin, is a harpoon of this class having two barbs.

Short Tang
Plate XXVI

This form differs from the last described in having a short shank and in seldom being more than three and one-half inches from point to point. They, like the long shank type, are usually cylindrical in section, and with a tapering tang, and are provided with a single barb. Occasionally the shank is notched on one or both sides.

Figure 4 shows a harpoon from Sheboygan County, three and one-half inches in length and three-sixteenths of an inch in thickness an inch from the point of the tang. It is rather flat in section and terminates in a sharp point, an inch above which a well-formed barb projects.

The Milwaukee Public Museum also possesses a series of five specimens of this type of harpoon, secured near Hancock, Michigan, which average from one and one-half to three inches in length.

A specimen of this form, having a notch in the shank, is in the State Historical Museum at Madison. The notch was doubtless for the attachment of a cord or line to be held in the hand when the shaft or handle was released, in order that the fish, when struck, might be played and controlled by the fisherman. This is the method in use at the present day in harpooning sword fish and other large and powerful inhabitants of the sea.

An example of this type, and of about the same length, formerly in the collection of W. H. Ellsworth, had two such notches opposite each other.

---

[67]Brown, Chas. E., Native Copper Harpoon Points, Wis. Archeologist, Vol. 7, No. 1, p. 51.

Fish Hooks
Plate XXVI

That the early inhabitants of this district depended to a considerable extent upon fish as food is evidenced by the fact that hundreds of copper fish hooks and scores of fragments of others have been recovered from the various village sites along the shore of Lake Michigan, the banks of the Wolf and Fox rivers, and many other locations where there must have been good fishing.

These hooks vary from one-half inch to six inches in length. They generally have the form of the modern fish hook, yet occasionally one is found that is square in section. The shank is usually straight, but sometimes notched, flattened or bent over to form an eye. The points are barbless and stand upward at various angles. Bits of sinew or twisted fiber, still attached to the shank, have been found in a few instances. Caches containing a number of fish hooks have been recovered in Wisconsin.[68]

The fact that the copper fish hooks have no barbs has led several archeologists to the conclusion that: "they must have been fastened to poles and used in hauling up fish where they gathered in shoals. By the exercise of sufficient skill and patience, of course, a solitary fish also might be taken with such a hook. Having no barbs, such hooks would often lose their burden".[69]

That the prehistoric Indian understood the use and advantages of the barb is indicated by the copper and bone harpoons recovered. That barbed fish hooks were unnecessary in their manner of fishing is quite certain. Several of the older Indians of Wisconsin, with whom the writer frequently fished, always waited for the fish to swallow the bait before setting the hook, in which case, the barb is of no advantage. Doubtless, much of their fish food was obtained by means of set lines, in the use of which the bait, containing the hook, was, as at present, invariably swallowed by the fish.

Figure 5 shows a hook two and one-half inches in length. The base is square in section and the point, as in many cases, turns outward to some extent. The shank is three-eighths of an inch in thickness.

---

[68]For additional information, see: Brown, Chas. E., in Moorhead, Warren K., Stone Age in North America, Vol. 2, p. 222.

Kuhm, Herbert W., Wisconsin Indian Fishing, Wis. Archeologist, new series, Vol. 7, No. 2.

[69]Winchell, N. H., The Aborigines of Minnesota, p. 500.

Figure 6 shows a specimen one and three-fourths inches long. The shank is flat and curved; the point doubtless bent out of shape.

Figure 7 shows a specimen one and one-fourth inches long, with a round shank.

Figure 8 shows a hook one and one-half inches in length. The shank is round with an exceedingly broad curve.

Figure 9 shows a specimen two and one-half inches in length. The shank is round with a very large hook.

Figure 10 illustrates a specimen three and seven-eighths inches in length. The shank is square and one-eighth of an inch in thickness.

Many of the short double-pointed spindles or awls may have been used as fish hooks. By fastening the line to the middle of the spindle, a sudden jerk, when the fish took the bait in which the hook was concealed, would serve to toggle the hook. This manner of fishing was practiced by historic Indians as toggles of the same type made of stone have been found in this state.

In the collection of Joseph Ringeisen, Jr., of Milwaukee, is an example with flattened shank, near the end of which is an eye for the attachment of the line.

Another example, three inches long, is in the collection of the Neville Public Museum, Green Bay, Wisconsin.

In the Dr. A. R. Whitman collection at Merrill, Wisconsin, is a stout fish hook, three inches long, which contains an eye.

A fish hook in the collection of Mr. Frank Miller, of Princeton, Wisconsin, collected within a few miles of that place, is one and one-half inches long. The end of the shank is turned to form a substantial eye and it is unique in having a copper staple passing through the loop or eye, one prong of which is seven-eighths of an inch long and the other one and one-quarter inches long. The point of the shorter prong is bent over the longer and cemented to it by the accumulation of verdigris, which accounts for its being in place. This specimen seems to indicate that the hook was fastened to a wooden float for surface fishing.

In the J. T. Reeder collection is a not uncommon example in which the end of the shank is slightly turned back, making the attachment of a line more secure.

CRESCENT-SHAPED OBJECTS

This class of copper objects is usually of crescent form, variously modified, with the addition of prongs or other prolongations arising from the ends or inner, upper edge. These objects called crescents, because of their shape, are usually referred to as ornaments and are supposed to have been worn suspended from the neck by means of cords. Few, however, contain perforations for such attachment, as do the copper plaques found in Ohio, and rarely in Wisconsin. There is, however, good reason to believe that most of them were not used as ornaments but as implements of utility.

Of the great numbers that have been collected in Wisconsin, most of them appear to have been obtained from ancient village sites, graves and burial mounds. An occasional one has been recovered in northern Michigan, Minnesota, Illinois and other parts of the country.

The outer curve of most types of crescents is bevelled to a sharp edge, while the inner was left rather blunt, leading to the conclusion that they were supplied with handles of bone or wood and used as chopping knives, scrapers, fleshers, or skin dressers, and for such other purposes as exigencies might demand.

These objects can be classed under three general types: prongless, canoe-shaped and pronged.

Prongless
Plate XXVII

This is not an uncommon form, and is usually found in all large collections. The blade of this type is broadly curved with outward sweeping extremities, ending in rounded tips. The width varies from three to nine inches. The outer or cutting edge is sharp, the inner one, blunt. This object could hardly be used without the addition of a handle. To make the attachment a groove was doubtless cut in the bone or wood and used for the insertion of the blade. A knife of this form, but having a slate blade is still in use by the Eskimo and is called "Woman's Knife".[70]

Figure 3 shows a specimen from Waupaca County which is purely crescent-shaped without projecting prongs. It is eight and three-eighths inches in length with the inner edge one-eighth of an inch in thickness. The outer rim is bevelled to a cutting edge, the width of the blade being one inch.

---

[70]Murdock, John, 9th Ann. Rept. Bur. Am. Ethnology, pp. 161-164.

Figure 6 shows a specimen from Marquette County which is nearly semi-circular. It is two and seven-eighths inches in length and belongs to the same type as the last above described. It has an outer cutting edge, while the inner edge is quite thick.

An example of this type secured from Monroe County, has both extremities notched.[71]

Another example of this type from Houghton, Michigan, six and one-quarter inches long and one and seven-eighths inches wide, is in the collection of Mr. J. T. Reeder of that place. This is a rather crude specimen and seems to be a modification of the general type in having the upper edge quite straight, the lower one being broadly curved. A number of specimens of this form with straight backs have been recovered.

A very small example of prongless crescent one inch long was collected near Hancock, Michigan, and is now in the Milwaukee Public Museum.

Another in the collection of Mr. Frank Miller, of Princeton, Wisconsin, is seven-eighths of an inch in length and contains a notch in the middle of the cutting edge by means of which a cord could be attached. It was found near Princeton.

Canoe-shaped
Plate XXVII

The profile of this type resembles that of an Indian canoe. The extremities of the blade curve inward and terminate in a short point or undeveloped prong. They were probably used with the addition of a handle to which they were attached in the same manner as that suggested for prongless crescents. This is the most common of crescent types and is represented in many collections.

Figure 4 shows a specimen from Fond du Lac County. It has an inner edge, one-eighth of an inch in thickness, which curves equally and terminates in a short point or prong, directed inward. The greatest width of the blade is one and one-half inches. The outer rim is bevelled to a sharp cutting edge.

A fine specimen of this type, from Calumet County, is in the Logan Museum at Beloit.

An example, from Door County, in the Neville Public Museum

[71]Brown, Charles E., Native Copper Ornaments of Wisconsin, Wis. Archeologist, Vol. 3, No. 2, p. 107.

at Green Bay, is unique, being in the shape of an ox-bow and having a decided projection from the inner edge.

The largest example known to the writer is from Oconto County and is now in the Hamilton collection at the State Historical Museum, Madison. The blade of this object is ten inches long and two and one-quarter inches wide. It weighs twenty ounces.

In this type of crescent, as well as in the prongless form, the outer curve and not the inner was beaten to a cutting edge, which leads to the conclusion that they were supplied with handles and used as implements of utility.

Pronged

The blades of this type have a crescent-shaped cutting edge but the back is usually straight with a blunted edge. A pair of prongs or spikes of various lengths projects from the upper part of the blade. These prongs or spikes may extend upward either from the outer extremities of the blade or from any part of the back between such extremities.

Pronged crescents can be sub-divided into three types: side prongs, center prongs and twisted prongs.

Side Prongs
Plate XXVII

This type of crescent has prongs extending upward at the extremities of the portion having the greatest width. The blade is beveled to a convex cutting edge and the back is straight, usually with a square edge. The prongs of this form of implement are merely continuations of the blade curved upward and were possibly used by being set into a handle. In other cases, the prongs were elongated and bent in the form of circles, large enough for the reception of two fingers in the use of the instrument.

Figure 2 shows a specimen from Shawano County. It is four and three-fourths inches long and the blade is one and one-half inches in width. The prongs, three-sixteenths of an inch in thickness, originally extended upward five inches, the ends being curved down, as shown in the figure.

Figure 5 shows a specimen from Waupaca County, of the same type as the last described. It is five and one-half inches in length, with projecting prongs, one inch in length, slightly curved inward.

The greatest width of the blade is one inch and the inner edge is one-eighth of an inch in thickness. The outer rim is bevelled to a sharp cutting edge.

A graceful example of this type is in the Bower collection at Minneapolis, Minnesota. The side prongs are flat and taper to points with an outward curve, making almost complete circles, through which two fingers could be inserted when in use. The blade of this instrument is bevelled to a cutting edge and the inner edge is blunt.

A specimen of this type with the extremities of the prongs united to form a small loop or eye was found in Marquette County, and was once in the collection of the late W. H. Ellsworth.[72]

Center Prongs
Plate XXVII

In this type the prongs leave the blade a fraction of an inch from each side of the center and extend upward. The outer edge of the blade is reduced to a cutting edge, while the inner side is usually straight and blunt. A handle could be easily attached between the prongs were it grooved to receive them.

Figure 8 shows a specimen from St. Croix County. It has a blade originally six inches in length, part of which is now missing. The outer edge is convex and worked very thin, with two prongs arising on either side of the middle of the base. Part of one prong is missing, the remaining one being five inches in length and three-eighths of an inch in thickness, two inches below the point of the prong.

A second example of this type, found in Texas, is in the Milwaukee Public Museum collection.

A somewhat broken specimen from Columbia County, the blade of which measures fully eight inches in length, is in the Logan Museum collection at Beloit.

A modification of this type, occasionally encountered, has a pair of spikes or prongs, usually rather long. They are either flat or cylindrical in section, and arise from either side of the middle of the top of the blade. The ends of the prongs are bent together and closely united by pounding, forming a bar.

Text figure 7 illustrates an example of this type from Waupaca

[72]Brown, C. E., Wis. Archeologist, Vol. 3, No. 3, pl. 16, fig. 122.

County, now in the collection of Mr. Joseph Ringeisen, Jr., of Milwaukee. The blade of this object is five inches long, and one inch wide. Three-fourths of an inch from either end there projects up-

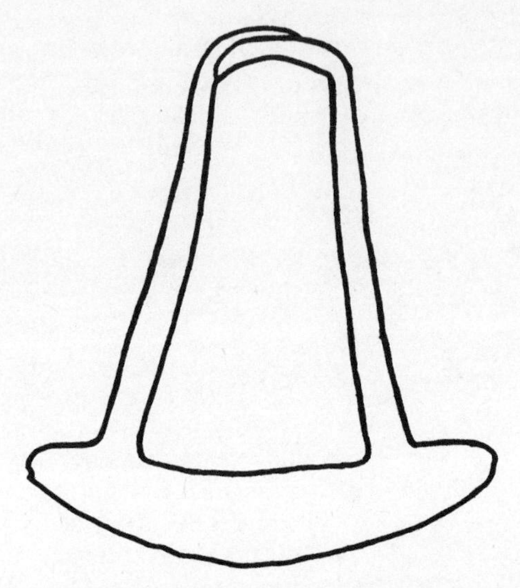

FIG. 7—Crescent-shaped copper object with center prongs.

ward from the inner part of the blade a prong, four inches in length. The ends of these are bent inward and united by annealing and pounding. These prongs are two and one-half inches apart at the blade and one and one-quarter inches apart where they turn to unite. These tapering prongs, like the blade, are beaten thin, the outer edge of the blade being sharp. While this implement may have been supplied with a handle, its thinness would indicate that it was not intended for hard service and as it could be suspended from the neck, it was probably used as an ornament.

A second example of the form last described, from Waushara County, is in the State Historical Museum collection at Madison. The crescent-shaped blade is six and one-half inches across and one and one-quarter inches in width. The tapering prongs are four and three-quarters inches in length, and about an inch apart. They are united at their tips by a cylindrical bar.

In the collection of Mr. Phillip C. Schupp, Jr., of Ravenswood, Illinois, are two examples of this type. The larger one, from Waukesha County, has a blade five and three-quarters inches long and one and one-half inches wide. The prongs are three and three-quarters inches in length and one and one-quarter inches apart. They rise from the middle of the upper side and turn at right angles toward each other forming a cross-bar two inches long.

The second was found in Winnebago County and has a blade three inches long. Two prongs three inches in length rise from the back and form a cross-bar at their extremities two and one-half inches long.

Twisted Prongs

In this form the prongs, which are beaten flat and thin, extend upward in a graceful curve, meeting a short distance from the blade, after which they are twisted and beaten, forming a tang, circular in section and gradually tapering to a point. This tang, from its shape, would seem to indicate the addition of a handle when in use.

Text figure 8 shows a splendid example of this type of crescent, found in Manitowoc County, and now in the cabinet of Mr. Joseph Ringeisen, Jr., of Milwaukee. The length of the blade is four and one-half inches and the width three-fourths of an inch. The length from the outer edge of the blade to the end of the tang is six and one-half inches. The blade and side prongs are flattened. The outer edge of the blade is reduced to a sharp cutting edge; its inner edge is blunt. The prongs come together two inches from the inner edge of the blade, and are twisted, pounded and annealed into a solid tang, tapering to a point.

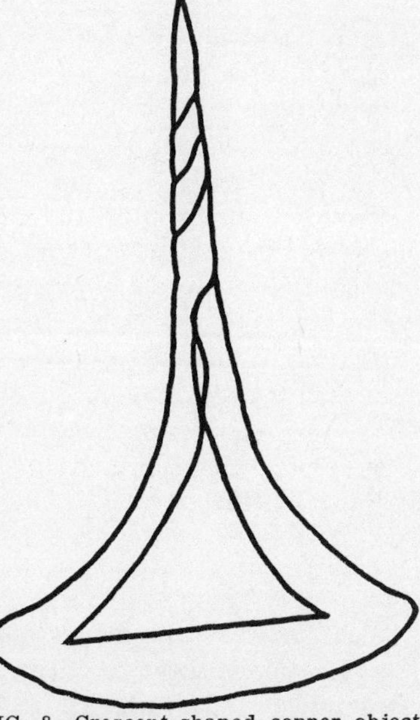

FIG. 8—Crescent-shaped copper object with twisted prongs.

A fine example, from Price County, is in the Hamilton collection in the State Historical Museum at Madison.

Another specimen of this form now in the Dr. Louis

Falge collection, was secured from an Indian grave in Manitowoc County.

In the S. D. Mitchell collection was one from Green Lake County, with part of one of the prongs broken.

A splendid example of this type is in the collection of Mr. Vetal Winn of Milwaukee and another is owned by Mr. J. F. Reeder, of Houghton, Michigan.

A number of others, some of them fragmentary, have been found in this state.

SPATULAS

Plate XXVII

Peculiarly-shaped implements, usually referred to as spatulas, have been recovered in small numbers in Wisconsin. The blades of these artifacts are thin, flat, broad and irregularly rounded. The tang is usually short, seldom more than one-fourth of an inch in thickness and either square or somewhat rectangular in section. It was doubtless intended for the attachment of a handle. The edge of the blade is usually sharp, indicating its use as a knife or scraper, although it may have been used for a diversity of purposes, including the scaling of fish.

Figure 1 shows a specimen from Shawano County. It is very thin with a total length of four and five-eighths inches and a greatest width of one and one-fourth inches. The blade, which is three inches in length, with a rounded point, is worked to a thin edge.

Figure 7 shows a specimen from Rock County, of a different form than the one previously described. It is four inches in length, with a rounded blade, two and one-fourth inches in width. The tang, one-eighth of an inch thick and two and one-half inches long, tapers to a rounded edge.

DECORATIVE OBJECTS

That a great proportion of the aboriginal copper ornaments, found in this district or taken from the mounds throughout the Ohio Valley and southward to the Gulf of Mexico, were made of thin sheets of copper is a fact beyond dispute.

Mr. J. B. McGuire, a well-known archeologist, believed that the sheet copper, from which these ornaments were fabricated, was of European origin. Mr. Clarence B. Moore, and other prominent archeologists, were of the opinion that these copper sheets were an American product, and that they were fabricated from nuggets of native copper by the aborigines. This led to a controversy between Messrs. Moore, McGuire, Putnam, Dorsey, Moorehead and Willoughby, as a result of which Mr. Moore conclusively showed that these articles were made of pure original native copper, differing notably in chemical composition from the European copper of the 16th, 17th and 18th centuries. This conclusion has been corroborated by Messrs. Moorehead, Putnam, Geo. A. Dorsey and C. C. Willoughby.

Mr. Moore had made several analyses of sheet-copper from mounds.

"Here is the result of an analysis of sheet-copper from Mt. Royal, made by A. R. Ledoux, M. S., PhD.:

Copper ....99.85 per cent
Silver ....trace
Iron ....trace

"Sheet-copper from the Grant mound, according to the analysis of Ledoux & Co., showed:

Copper ....99.730 per cent
Iron ....00.34 per cent
Silver ....00.023 per cent"

These analyses were compared with those of nuggets of copper from the Lake Superior region, which showed the only impurities to be traces of silver and of iron. They were also compared with analyses he had made from sheets of copper found on the surface, and admitted to be of European origin, as well as portions of brass and copper trade kettles, which in every case, showed the presence of arsenic, antimony, zinc, and sometimes nickel and bismuth. As none of these impurities were found in our native copper, the conclusion was that the aborigines were able to and did produce sheets of this metal. One argument against it was the difficulty of pounding out a nugget, as it was found to crumble during the hammering process, and continued to do so whether it was hammered cold or hot, but experiments by Mr. Charles C. Willoughby proved that by careful hammering, annealing, grinding, cutting, embossing, and polishing, a nugget could be made into a thin sheet. Mr. Willoughby demonstrated this successfully by taking a nugget from an altar of an Ohio mound and a second one of native copper from the Lake Superior region. From each, a thin sheet of copper was successfully made, the process being described by Mr. Willoughy as follows:—

"The experiment was carried out upon a sea-beach strewn with water-worn stones of all sizes. Placing upon a smooth stone a piece of native copper from the Lake Superior region, and using an oval water-worn stone as a hammer, the copper was carefully beaten. A few blows sufficed to show the tendency of the copper to crack along the edges as it expanded. This tendency was overcome by annealing. When the sheet had attained the required size it was ground to a uniform thickness between two flat stones, the work being hastened by the addition of fine sand."[78]

A very small percentage of the copper artifacts found in the Lake Superior District can be termed ornaments. Their classes and types are not large, nor are they artistic in form, being confined mainly to beads. It is to be remembered, however, that the beads and ornaments of collections, which to us appear dull and crude, when in the possession of the primitive Indian, were highly polished and attractive.

Beads and Bangles
Plate XXIX

The usual Wisconsin type is somewhat spherical in shape and made of thin strips of copper rolled around a spindle or like object, the number of turns regulating the size of the bead. They were then smoothed and polished. In some cases solid pieces of copper were drilled from each side and the exterior shaped afterward. In other instances, the strip of metal was bevelled on both edges before being rolled in order to make the beads spherical.

Figure 8 shows one of a number of strings of beads of different types and sizes owned by the Milwaukee Public Museum. This one, from Lyman County, South Dakota, is eleven feet six inches in length and contains five hundred and sixty copper beads. Each

---

[78]Jour. Acad. Nat. Sci., Philadelphia, Vol. X, 1894. Certain Sand Mounds of the St. John's river, Florida, Part II, p. 123; Am. Anthropologist, new series, Vol. V., pp. 27-57, 1903; Discussion as to Copper From the Mounds; a reprint of articles by Moore, McClure, Putnam, Dorsey, Moorehead, and Willoughby, Am. Anthropologist, Vol. 5, No. 1, Jan'y-March, 1903.

was hammered smoothly into shape after rolling and each is one-fourth of an inch in diameter.

Figure 5 shows a second string of beads, from Wisconsin, owned by this Museum. It consists of sixty-one heavy rolled beads, fifty-five of which graduate from three-eighths to one-eighth of an inch in diameter. Three of the others average from one and one-fourth to three and one-fourth inches in length, and the remaining three are from one-fourth to one-half inch in length.

Figure 7 shows an example of an oval bead, one-half inch in diameter, which is drilled from each side.

Figure 1 represents a bead of rolled copper, tubular in form, and three-fourths of an inch in length.

Occasionally a bead is found which is made from a piece of sheet copper, rolled from each end, thus making two connected tubes. Figure 6 shows an example one inch in length.

Figure 2 shows a specimen of rolled copper, conical in shape, and seven-eighths of an inch in length. It was used either as a bead or a bangle.

Bangles made in this form, but usually perforated at the small end, were frequently used for the decoration of garments.

Bracelets
Plate XXIX

Bracelets are commonly found in the central southern states, but Wisconsin finds are few. Specimens, recovered in this district, vary considerably in form. The larger ones may have been used as anklets or armlets.

Trade bracelets of copper, silver, and brass have been recovered here in considerable numbers.

Figure 4 illustrates a type of historic copper bracelet, fifty of which are in the Milwaukee Museum collection.

The prehistoric bracelets which are reported as collected, are generally irregular circlets, about one-fourth of an inch in thickness, the ends usually coming close together. In some instances, the ends are turned back to form small loops. A few consisting of a simple coil of native copper have been recovered from mounds and graves. Sheet copper bracelets have been reported as found in several parts of Wisconsin, but none have come under the writer's observation.

Finger Rings
Plate XXIX

Finger rings of native copper are seldom found here.

Figure 3 shows a type of copper ring which consists of a small, narrow, rounded strip of native metal bent into the form of a simple circlet, the ends nearly meeting. The strip, from which this example was made, was seven-sixteenths of an inch in width. A few specimens have been secured which are thickest at the middle, and taper to a point at the extremities. It is quite likely that many of these so-called rings were suspended from the ears when in use.

A second form consists of coiled copper wire and a few are made of narrow strips of sheet copper, the ends of which are brought close together, but are never joined or over-lapped.

Gorgets

Gorgets of copper are exceedingly rare in Wisconsin. The most interesting example coming to the writer's notice is in the collection of Mr. Joseph Ringeisen, Jr., of Milwaukee, and is shown in text figure 9. It is of beaten copper and is three and one-fourth inches long, with rounded ends and concave sides. It is three-fourths of an inch wide at the middle, and three-eighths of an inch thick. The ends are nearly one and one-half inches wide and are drawn quite thin. Near the edge of one end there is a small perforation, evidently punched and not drilled.

FIG. 9—Copper gorget.

A more common type of gorget is circular in form. It is made of beaten copper and has one or two perforations, usually near the edge, but occasionally through the center.

An exceptional example from Kenosha County, formerly in the W. H. Ellsworth collection, at Milwaukee, was three and one-fourth inches in diameter and one-sixteenth of an inch in thickness.

This specimen was beautifully oxidized and one side was worn smooth.

FIG. 10 — Copper pendant.

Disks of this type with a single perforation are reported from a dozen or more Wisconsin counties.

In the S. D. Mitchell collection there is a small copper disk, from Green Lake County. It is one and one-fourth inches in diameter and has notched or scalloped edges, and a small perforation at the center.

In the State Historical Museum collection, at Madison, Wisconsin, is an example, nearly triangular in shape, with a single perforation near the apex of the upper angle. The base is two and three-fourths inches wide and the corners turn down forming short prongs.

Pendants

Pendants of copper are closely allied to the gorgets, each being provided with one or more perforations for the reception of a cord to enable the object to be suspended from the neck. Those made by beating the metal to a thin sheet are usually classed as pendants. The number found in this district, being small, leads to the suggestion that, because of their thinness many may have disappeared as the result of oxidization. Objects of this type, collected here, are not embossed or ornamented as are those found in much greater numbers in the southern and central southern states.

Text figure 10 illustrates a fine specimen from Green Lake County, formerly in the W. H. Ellsworth collection, at Milwaukee. It was nearly triangular in shape, five and three-fourths inches long and two inches wide at the widest part. It had two small perforations near the upper edge and was only one-sixteenth of an inch in thickness. A few pendants of this type have been found in this district.

In the J. G. Pickett collection there is a leaf-shaped example, from Winnebago County, which is three and one-half inches in length and one and one-quarter inches wide at its widest portion. It has two perforations near the upper edge.

In the J. T. Reeder collection are two specimens of this type from Houghton County, Michigan.

A specimen, from Manitowoc County, in the collection of Mr. Vetal Winn, of Milwaukee, is a plate of copper, about one and one-half inches long and one inch wide. It is nearly triangular, the top being convex and having a perforation and two indentations on the face.

In the collection of Mr. Frank Miller, of Princeton, is a specimen, secured near that place, that is nearly circular. It is one and one-half inches in width, and has, near one edge, a perforation one-half inch in diameter.

In the collection of Mr. Phillip C. Schupp, Jr., of Ravenswood, Illinois, is a fine example from Winnebago County, three inches long, and two and five-eighths inches wide. It is made of thin copper containing large patches of silver, and has two perforations near one side for the attachment of a cord.

Others are reported from Brown, Barron, Crawford, and Jefferson counties.

Hair Ornaments

One type of these ornaments has a blade nearly circular in form, with the ends terminating in two long parallel prongs. This object is flat in section. The prongs are very close together and indicate its use as a hair ornament.

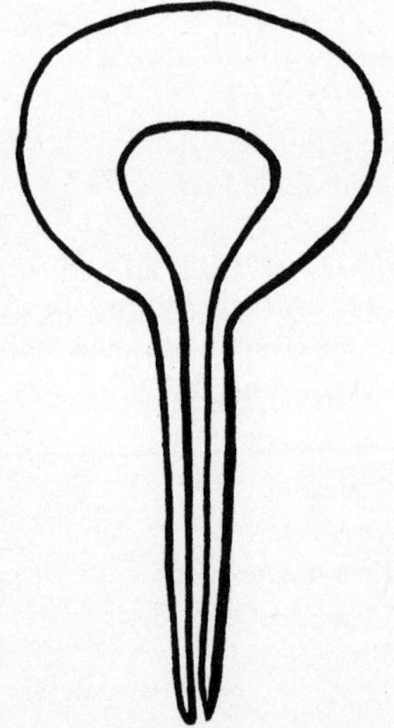

FIG. 11—Copper hair ornament of unusual form.

Text figure 11 shows a specimen of this unusual form from Roseau County, Minnesota, now in the cabinet of Mr. P. O. Fryklund, of Roseau. The blade of this ornament is three inches long and nearly one inch wide. It is almost circular and terminates in two parallel, flat, pointed prongs, each three and one-half inches in length. These prongs are one-fourth of an inch apart. The greatest thickness of any part of this object is three-sixteenths of an inch. The shape would indicate its use as a hair ornament.

Some other crescent forms may have been used for the same purpose. Copper rods, tapering to a point at each end have been mentioned as hair ornaments.

In the Milwaukee Public Museum are two flat strips of copper, from Sheboygan County, ornamented along each edge and down the center with a row of deep indentations. One is six inches long and one inch wide, and the other is somewhat shorter.

Ear Spools

Ear spools, made of thin sheet copper, have been obtained in considerable numbers from the Hopewell Mound Group, Ohio, other localities in that state, and in some of the southern states. A few have been recovered in Illinois. Some of these are nearly two inches in diameter, and are composed of two cones of hammered copper which are joined together at their apices, and quite elaborately ornamented with embossed figures. That they were worn by being suspended, probably from the ears, is evidenced from the fact that a few have been found with the remains of a cord attached. It is supposed that these objects were also attached by means of rivets.

Prof. T. H. Lewis reported that he secured ornaments of this nature from mounds in Crawford and Grant counties, Wisconsin.

A second well-known form, shown in text figure 12, is made of thin beaten native copper. It is spool-shaped and ornamented, and suggests two short spouted funnels, with the ends of the spouts united.

FIG. 12—Copper ear spools.

Another type of ear ornament or ear plug, which is much smaller, consists of a rounded section of wood or bone, covered with a thin sheet of copper or silver. Examples of this type were recovered by members of the Milwaukee Public Museum expedition, while excavating ancient Aztalan, Jefferson County.

Plaques

These objects are extremely rare in Wisconsin. Those recovered are from mounds and are associated with burials. They are usually rectangular, of various sizes, and are made of native copper, beaten into a thin sheet. They have two or more small perforations, evidently for the reception of a cord, by which they could be suspended from the neck or fastened to some other part of the body.

The first record of the discovery of artifacts of this type in Wisconsin seems to be by Dr. Cyrus Thomas who excavated mound No. 6, in White's Group, Crawford County. This mound contained twelve badly decomposed skeletons. On the skull of one of these was found a thick copper plate or plaque, "apparently beaten out of native copper with rude implements". It was eight inches long and four and one-half inches wide. It had two perforations near the center of the upper edge and a double row of circular indentations extending across the face near each end.

The second copper plate, also secured by Dr. Thomas, was about four and one-fourth inches square.[74]

Two beaten native copper plaques, without ornamentation, were recovered, during the past year, from mounds in Trempealeau County, by a Milwaukee Public Museum expedition, under the direction of W. C. McKern, Associate Curator of Anthropology, who ascribes them to the Ohio Hopewell culture. One of these plaques is about eight inches long by five inches wide and has two perforations near the middle of one long edge. On the under side are adhering pieces of cloth or woven fabric, probably preserved through contact with the copper. The second plaque is six inches long and two and one-half inches wide. Both were associated with artifacts of the Ohio mound culture type.

CEREMONIALS

Plate XXX

Bird stones, quite commonly found in many parts of the United States, are of unknown use but it is supposed that these objects were merely amulets. None made of copper have been reported. The object, shown in figure 1, may possibly belong to the bird stone class. It is made of copper, beaten into the shape of what appears to be intended for a deer. The body of this specimen is one and one-third inches in length, including the head, and one

[74]Thomas, Cyrus, Mound Explorations, 12th Ann. Rept. Bur. Amer. Ethn., pp. 80-81, Wash., 1890-91.

and one-half inches in height. Through the upper part of the body is a perforation, probably for the reception of a cord, in order that it might be worn suspended from the neck. Near the extremities of the lower part of the body are two other holes that extend from side to side. This interesting object belongs to Mr. Joseph Ringeisen, Jr., of Milwaukee, and is unique in being the only one of its kind reported.

Figure 2 shows a specimen which may be classed as a copper banner stone. It is shaped somewhat like an ancient two-bladed battle axe, and probably was used for ceremonial purposes. The median length is five inches. The blades are five inches in width and each is hammered to a very thin edge. The center is one inch in thickness, and is pierced by an eye or hole, one-fourth of an inch in diameter, probably for the reception of a handle. This remarkable object was found at Fond du Lac, Wisconsin on August 1, 1916, some six feet below the surface, by James McCabe, a local contractor, while constructing a sewer, and is now the property of Mr. Joseph Ringeisen, Jr., of Milwaukee.

A careful examination of this specimen discloses the fact that it is formed of two separate plates of copper, annealed and welded together. The line of union can be traced completely around the object. The eye or hole passing through the center was not drilled, but was formed by hammering around a core.

Ceremonial axes are quite common in the central southern states and in other parts of the United States. They are so-called, because of the general opinion that they were supplied with short wooden handles and carried by persons in authority on ceremonial occasions. They vary considerably in shape and size, and usually are fashioned from slate, porphyry, granite, or other material pleasing to the eye. There are two copper specimens of this class in the H. P. Hamilton collection, at the State Historical Museum at Madison.

"One of these, found at Oconto, Oconto County, Oct. 1, 1899, was included in a remarkable cache of copper implements and ornaments, consisting of a crescent, sword, chisel, leaf-shaped blade and arrow points. This specimen, weighing five ounces, is three and one-half inches in length, and one and one-fourth inches in width across the elevated part at the middle. The broad wings are one and one-fourth inches in length and one and one-half inches in width across their outer edges. The perforation at the middle has a long diameter of one inch and a short diameter of half an inch. A second specimen in the same cabinet, weighing two and one-fourth ounces, comes from the same locality. It is five inches in length and only one inch in width across the widest part, near the middle. The narrow wings are two and one-fourth inches in length and taper to a rounded point, the perforation at the middle being half an inch in diameter.

"In the collection of Dr. H. M. Whelpley, St. Louis, Mo., among a remarkably fine series of hematite axes, celts, hammers, pendants, gorgets, pipes, cones, hemispheres, and other objects, are two hematite banner stones. Others are in the Louisiana State collection now on exhibition at St. Louis, Mo. These, and the Hamilton specimens, are the only metal banner stones at present known to the author."[75]

CACHES

Caches of native copper implements have been quite frequently found in Wisconsin and are often associated with partly finished implements and copper fragments, which is further evidence that these implements were manufactured here.

Among the caches discovered is one from Calumet County now in the Milwaukee Public Museum collection. It consists of four spear points. Another in the State Historical collection consists of a copper knife, three spear points and five perforators, found on the Two Rivers' village site; a second one, consisting of two small arrow points, a leaf-shaped blade, axe, banner ceremonial crescent, and a large curved knife came from Oconto County; and a third one consisting of an axe and two harpoons from Outagamie County.

Another known as the Fond du Lac cache, from Fond du Lac County, consists of twenty-one pieces, of which sixteen were spear points and three, pikes. An awl, and a cylindrical implement, six inches long and one inch in diameter, made up the rest of the cache. One end of the cylindrical implement is bored to a depth of two inches which makes that extremity a tube while the remaining portion is of solid metal.

In the State Historical Museum at Madison is a cache of ten fish hooks, from Waupaca County.

---

[75]Brown, Chas. E., The Native Copper Ornaments of Wisconsin, Wis. Archeologist, Vol. 3, No. 3, p. 121.

"John Jerend at one time found fifty fish hooks of various sizes that had just been exposed by the wind." These came from the Black River village site, Sheboygan County.

It is also reported that "Mr. Rudolph Kuehne, of Sheboygan, obtained from the New Amsterdam site, six copper fish hooks. These lay on top of each other and had become united in a mass of verdigris. All are of slender form with a notched shank. Four other fish hooks lay nearby".[76]

In 1920 a cache of eight chisel-like implements and a dozen shapeless fragments of copper were found on the bank of Pelican River, Oneida County, and later two more chisels were uncovered by the plow in the same place. With one exception, all the pieces of the cache were found in a space one rod in diameter, the remaining one evidently being carried by the plow for a short distance. Three of the largest of these implements are now in the collection of Mr. Phillip Schupp, Jr., of Ravenswood, Illinois, and are described in this paper under the title of Adzes.

Five conical spear points were found on a ledge in Door County.

None of the objects, comprising these caches, show any evidence of use.

FRAUDULENT SPECIMENS

While the traffic in fraudulent material in America is not so extensive as to put collectors constantly on their guard, it is the more dangerous, especially to beginners in archeology and those of wealth who do not exercise care in their purchases.

Copper implements have been fabricated by unscrupulous white men, by pouring melted copper or brass into molds or forms made from genuine specimens. Some of these fraudulent productions are known to have been rejected and condemned by several collectors, and yet have found a resting place in the cabinet of some unsuspecting man of means. These counterfeits have all the marks, elevations and depressions of the original, and, after being treated with acids, possess the characteristic patina or green coating that aids in deceiving the inexperienced and unobserving collector. Such a coating is thin, never incrusted, and can be easily removed.

Other frauds in copper are occasionally encountered. Some are cut from heavy sheet copper, in the form of an arrow or spear point.

[76]Kuhm, Herbert W., Wisconsin Indian Fishing—Primitive and Modern. Wis. Archeologist, new series, Vol. 7, No. 2.

These are perfectly flat and while showing the green coating have none of the characteristic evidences of erosion that the genuine objects show.

In the genuine native copper implement, the grain of the metal, by reason of being drawn out, runs lengthwise of the object. In cast implements of this sort, it has been found that the grain of the metal crosses the object, which can be determined by the application of a strong acid.

Another variety of fraud is the pernicious practice of a few dealers who sell specimens with the declaration that they come from some desired locality, when in fact they were not found there. This practice should also be condemned, its effects being equally as contaminating as if the specimens themselves were fraudulent.

TABULATION OF NATIVE COPPER OBJECTS
in the
MILWAUKEE PUBLIC MUSEUM

Category	Type	Subtype	Number
SPEAR and ARROW-POINTS	Eyed Tang		6
	Spatulate Tang		103
	Notched Tang		5
	Toothed Tang		26
	Socketed Tang		335
	Conical Points		42
	Long Rat-tailed Tang		31
	Short Rat-tailed Tang		53
KNIVES	Curved Back		17
	Straight Back		66
	Handled		11
	Socketed		2
	Double-edged		20
CELTS	Spuds	Long Socketed	20
		Short Socketed	10
	Axes	Flat Blade	22
		Bell-shaped	2
		Grooved and Notched	1
		Double-bitted	2
	Chisels	Broad Cutting Edge	41
		Uniform Width	7
		Median Ridge	4
ADZES			3
WEDGES			2
GOUGES	Expanded Cutting Edge		2
	Uniform Width		2

Category	Type	Subtype	Number
PIERCING IMPLEMENTS	Pikes		20
	Punches		3
	Drills		21
	Awls		102
	Needles		136
FISHING IMPLEMENTS	Harpoons	Multiple-barbed	1
		Socketed Tang	2
		Long Tang	8
		Short Tang	8
	Fish Hooks		69
CRESCENTS	Prongless		7
	Canoe-shaped		4
	Pronged	Side Prongs	13
		Center Prongs	5
		Twisted Prongs	0
SPATULAS			4
DECORATIVE OBJECTS	Beads		1043
	Bracelets	(Historic)	50
	Rings	(Historic)	2
	Gorgets		0
	Pendants		6
	Hair Ornaments		0
	Ear Spools		0
	Plaques		3
	Ceremonials		0
UNCLASSIFIED and WORKED PIECES			252

EXPLANATION OF PLATE I.

Commander E..F. McDonald, Jr.

EXPLANATION OF PLATE II.

Mr. Burt A. Massee.

EXPLANATION OF PLATE III.

Associate Members of the Mc-Donald-Massee 1928 Isle Royale Expedition.

Figure 1. Mr. Geo. R. Fox.
Figure 2. Prof. Baker Brownell.
Figure 3. Dr. Alvin LaForge.
Figure 4. Mr. Geo. A. West.
Figure 5. Mr. John Kellogg.
Figure 6. Mr. Jos. T. Reilly.
Figure 7. Mr. John F. Dunphy.

1
2
3
4
5
6
7

EXPLANATION OF PLATE IV.

Figure 1. Yacht "Naroca".

Photograph by Geo. R. Fox.

Figure 2. Yacht "Margo".

Photograph by Geo. A. West.

1
2

EXPLANATION OF PLATE V.

Figure 1. Beautiful Mackinac.

Figure 2. Historic Sault Ste. Marie. Yachts passing through the locks.

Photographs by Geo. A. West.

1
2

EXPLANATION OF PLATE VI.

Figure 1. Rock Harbor and Rock Harbor Lodge.
Photograph by Geo. A. West.

Figure 2. "Susan Cave" near Rock Harbor.
Photograph by Geo. R. Fox.

EXPLANATION OF PLATE VII.

Figure 1. Lighthouse—Rock Harbor.

Photograph by Geo. R. Fox.

Figure 2. Richie Lake.

Photograph by Geo. A. West.

1
2

EXPLANATION OF PLATE VIII.

Figure 1. "Massee Rock Shelter"—Point Houghton. An ossuary from which was taken a number of skeletons.

Figure 2. Fishermen's Home—Point Houghton.

Photographs by Geo. R. Fox.

1
2

EXPLANATION OF PLATE IX.

Skulls from ossuary, "Massee Rock Shelter", recovered by the 1928 Expedition.

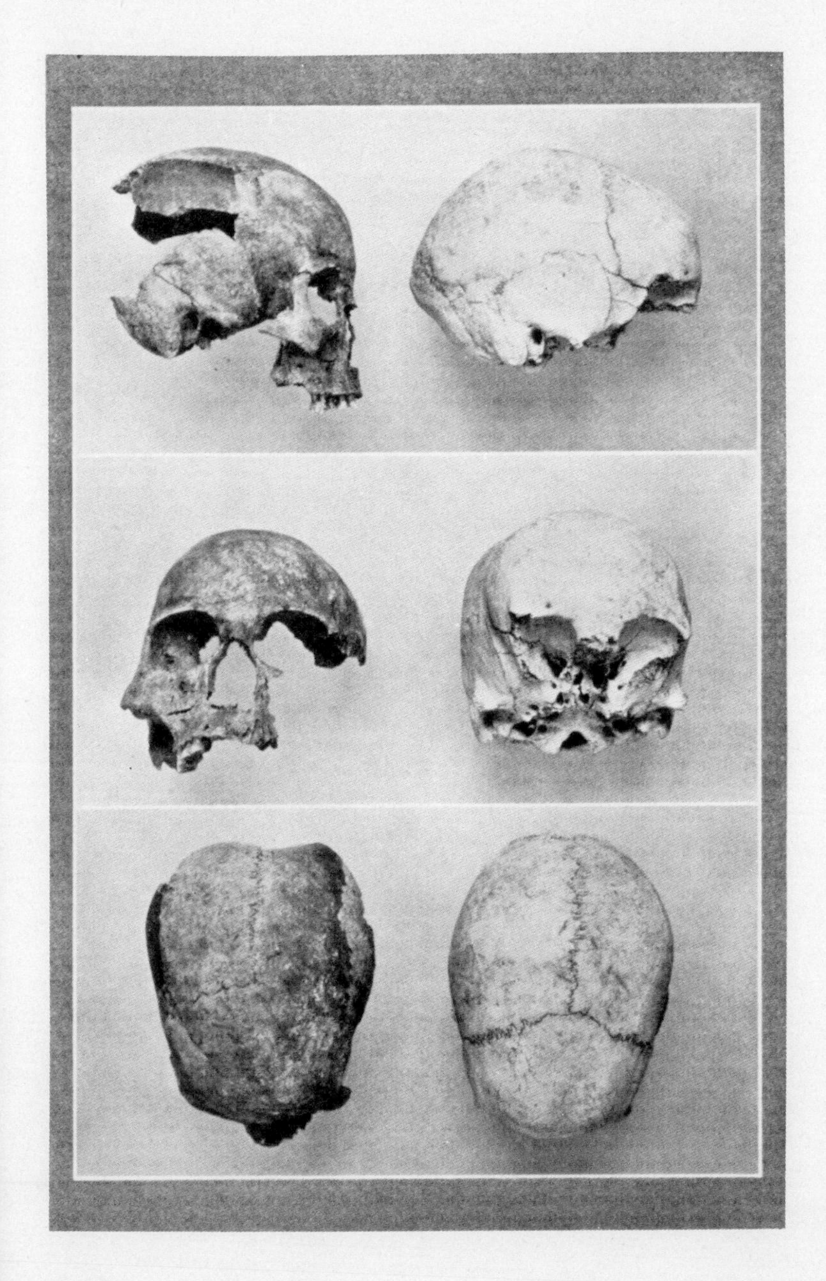

EXPLANATION OF PLATE X.

Figure 1. "Old Mine" workings in the interior, three miles from Hay Bay.

Figure 2. Excavated pit in the "Old Town", near Hay Bay.

Photographs by Geo. R. Fox.

1
2

EXPLANATION OF PLATE XI.

Figure 1. Ancient beach line near Hay Bay.

Figure 2. Ferguson's "Pit Dwelling", "Old Town", near Hay Bay.

Photographs by Geo. R. Fox.

1
2

EXPLANATION OF PLATE XII.

Figure 1. Mine beneath trap rock.

Figure 2. Tree within the "Old Fort".

Photographs by Geo. R. Fox.

1

2

EXPLANATION OF PLATE XIII.

Figure 1. Swimming Moose—Siskowit Lake.
Photograph by Geo. A. West

Figure 2. Village Site—Chippewa Harbor.
Photograph by Geo. R. Fox.

1
2

EXPLANATION OF PLATE XIV.

Figure 1. Log dwelling two stories high, Chippewa Harbor.
Photograph by Geo. A. West.

Figure 2. Birch Island at entrance to McCargoe Cove.

1
2

EXPLANATION OF PLATE XV.

Figure 1. Mr. and Mrs. John Linklater, whose summer home is Birch Island.

Photograph by Geo. R. Fox.

Figure 2. McCargoe Cove—The yacht, "Naroca", and the schooner, "Swastika".

Photograph by Geo. A. West.

1
2

EXPLANATION OF PLATE XVI.

Figure 1. Pit at McCargoe Cove excavated by the Milwaukee Public Museum Expedition of 1924.

Figure 2. Mauls and hammer-stones, McCargoe Cove.

1
2

EXPLANATION OF PLATE XVII.

Isle Royale Potsherds.

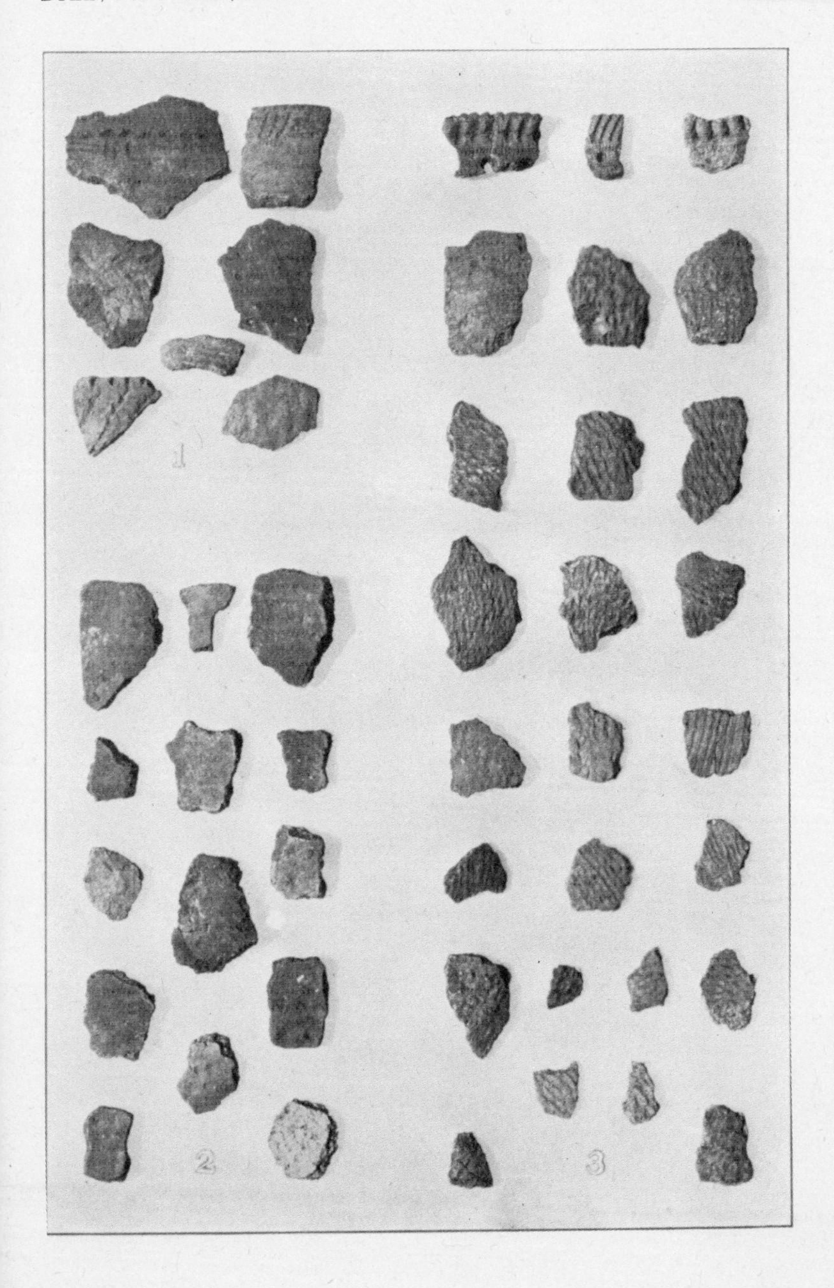
1
2
3

EXPLANATION OF PLATE XVIII.

Isle Royale Stone and Copper Artifacts.

Figure 1. Obsidian arrow point found on the shore of Siskowit Lake.

Figure 2. Chalcedony arrow point found within Ferguson's "Old Fort".

Figure 3. Chert knife found in the ossuary at Point Houghton.

Figure 4. Copper implement found at McCargoe Cove.

Figures 5 and 6. Chert scrapers found on Chippewa Harbor village site.

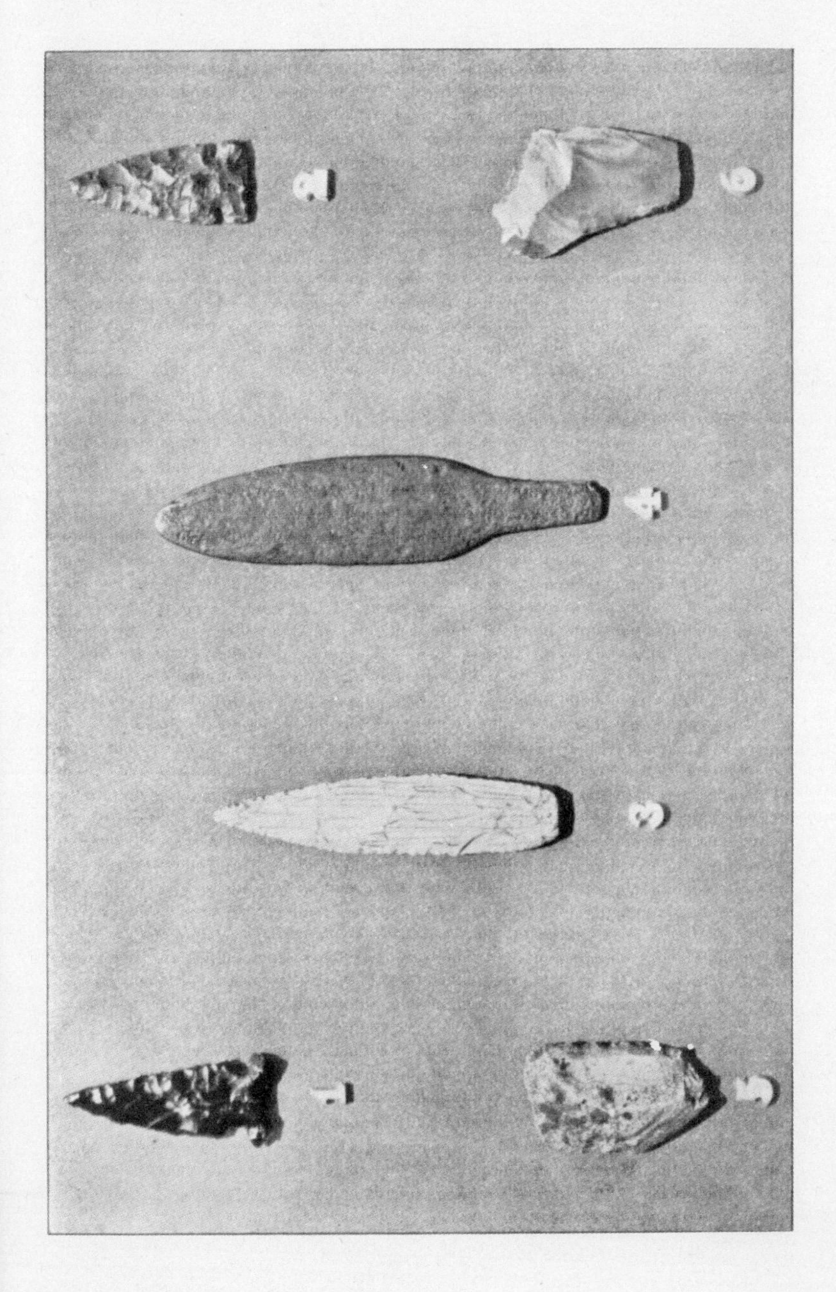

EXPLANATION OF PLATE XIX.

Figure 1. Stone implements secured on Isle Royale by Mr. F. M. Warren of Minneapolis.

Photograph by F. M. Warren.

Figure 2. Tree growing on dump of ancient mine, McCargoe Cove.

1

2

EXPLANATION OF PLATE XX.

Types of copper spear and arrow-points and their modifications—Milwaukee Public Museum collection.

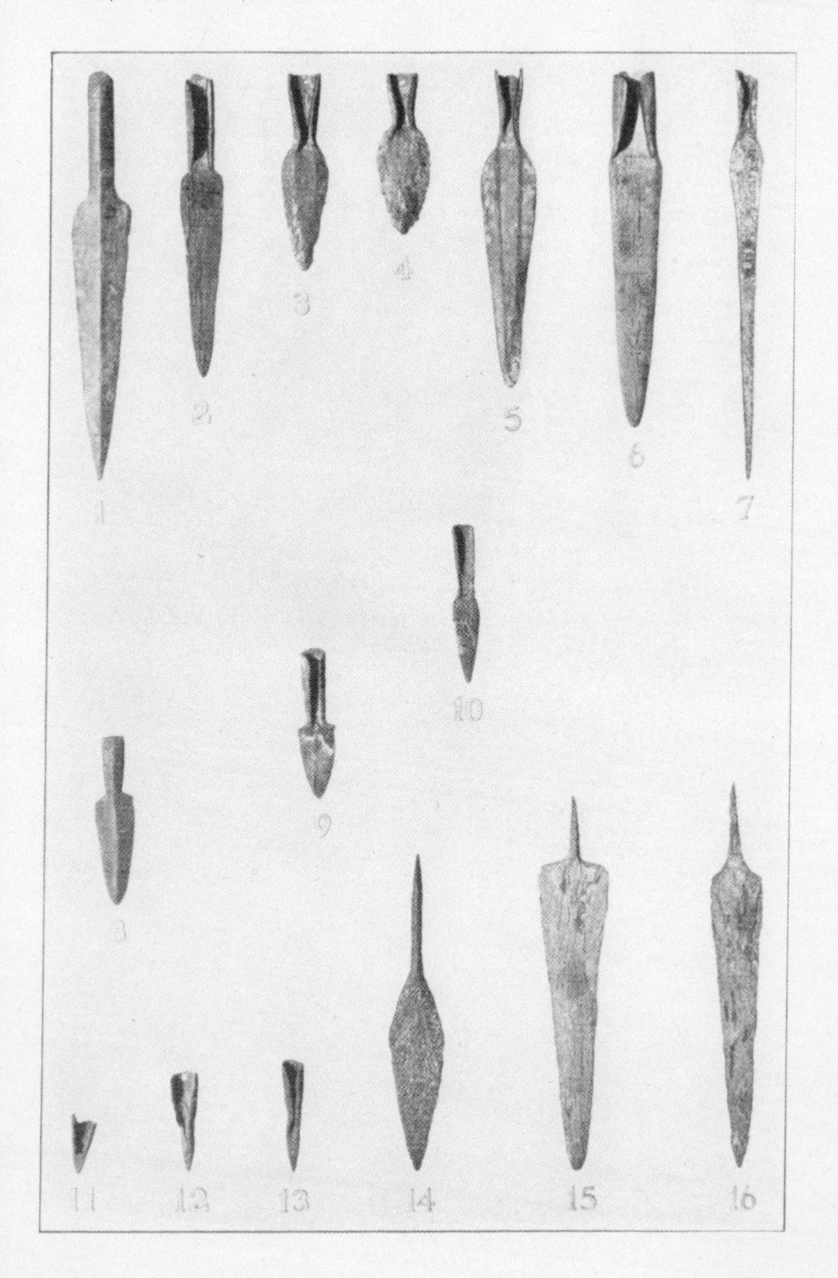
1
2
3
4
5
6
7
8
9
10
11
12
13
14
15
16

EXPLANATION OF PLATE XXI.

Types of copper spear-points and their modifications—Milwaukee Public Museum collection.

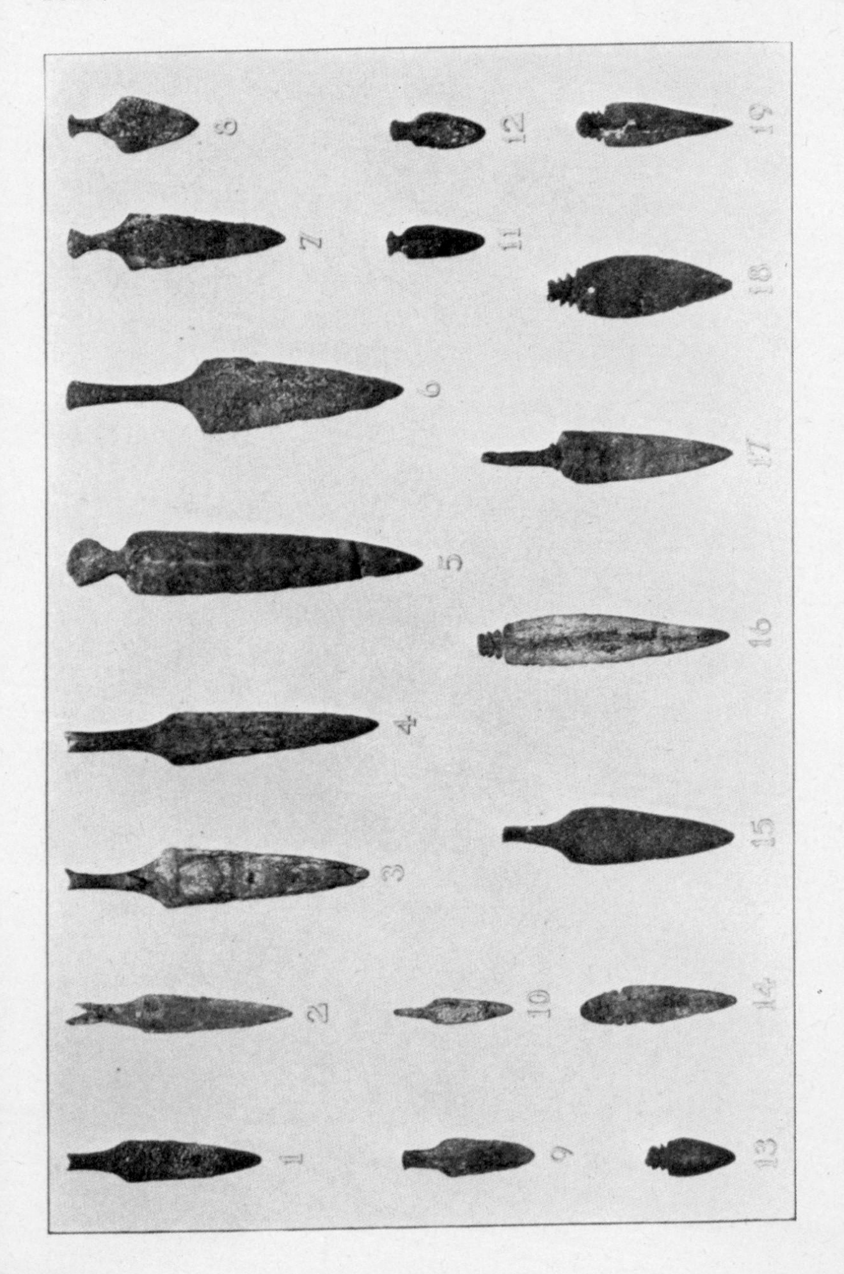
1
2
3
4
5
6
7
8
9
10
11
12
13
14
15
16
17
18
19

EXPLANATION OF PLATE XXII.

Types of copper knives and their modifications—Milwaukee Public Museum collection.

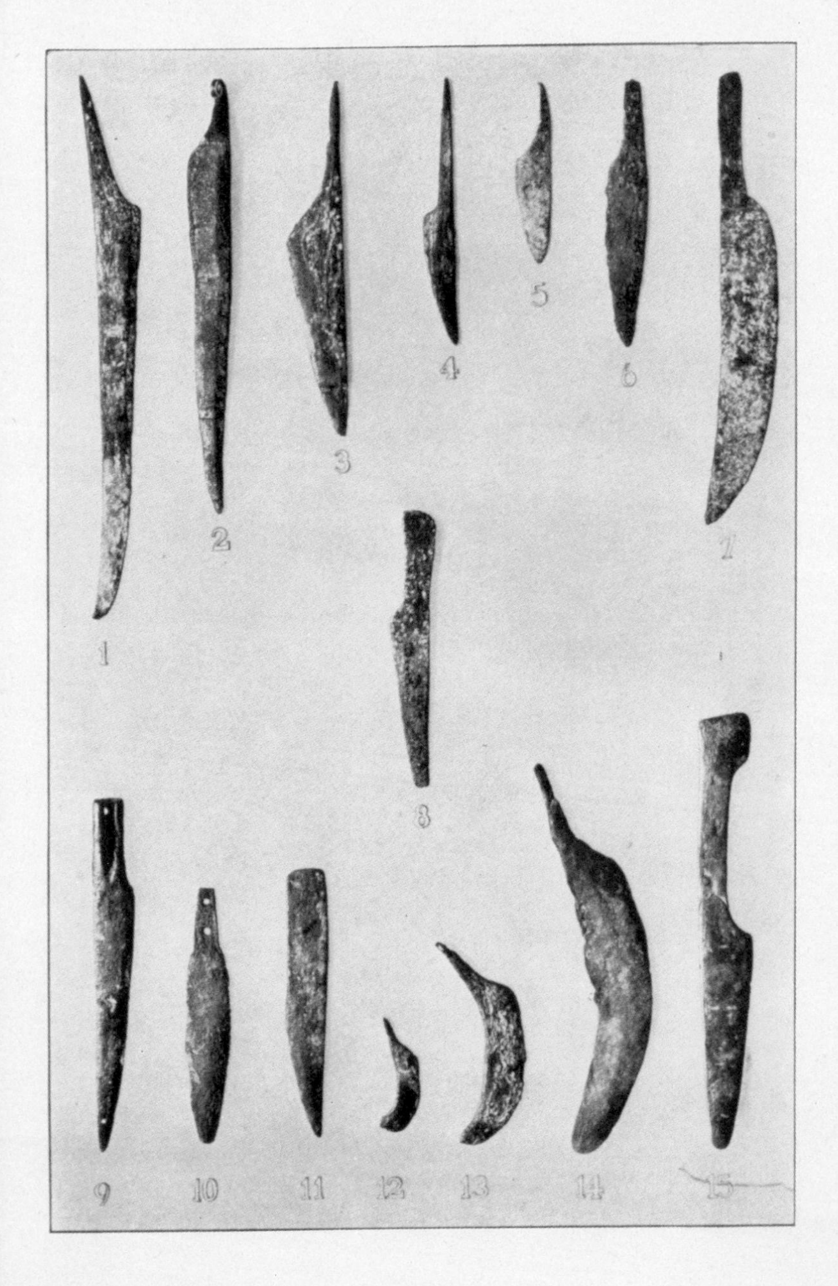
1
2
3
4
5
6
7
8
9
10
11
12
13
14
15

EXPLANATION OF PLATE XXIII.

Types of copper spuds, wedges and gouges—Milwaukee Public Museum collection.

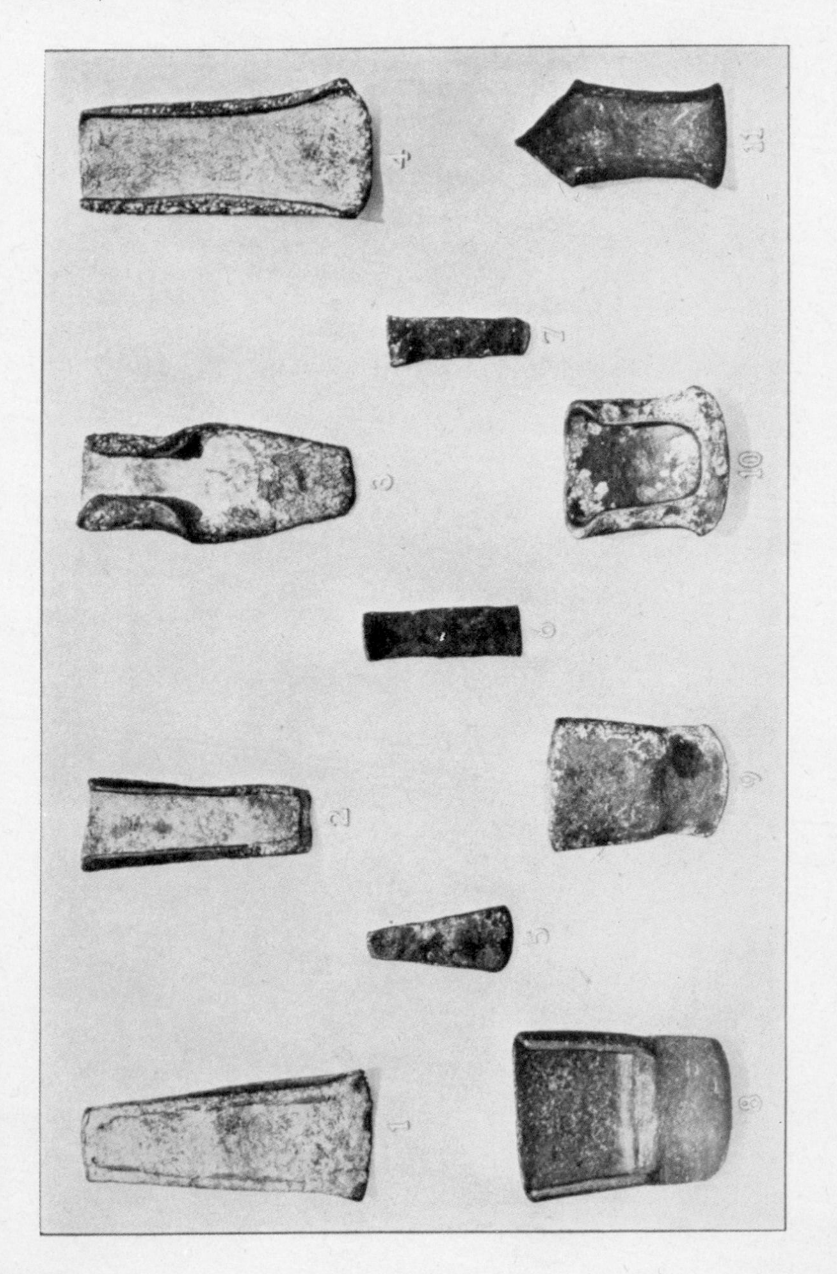

1
2
3
4
5
6
7
8
9
10
11

EXPLANATION OF PLATE XXIV.

Types of copper chisels and axes—Milwaukee Public Museum collection.

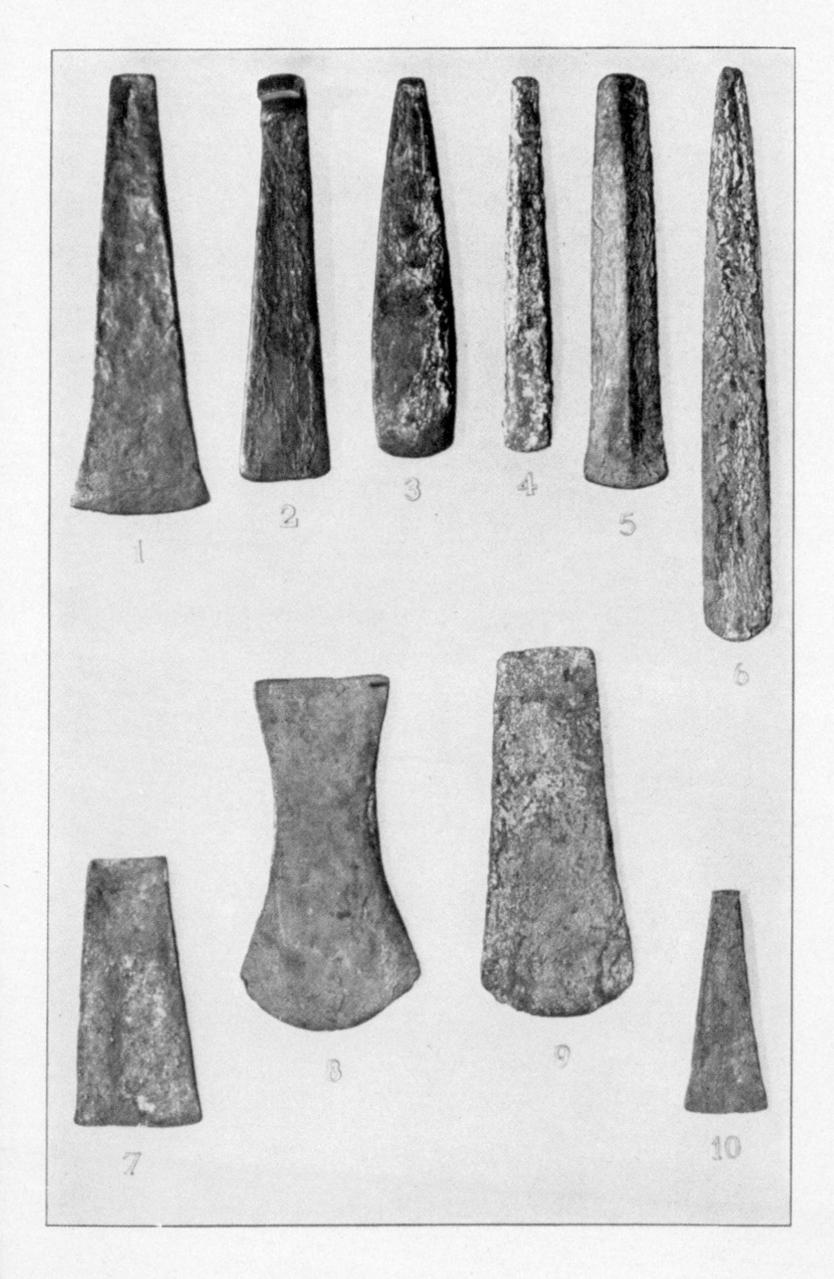
1
2
3
4
5
6
7
8
9
10

EXPLANATION OF PLATE XXV.

Types of copper pikes, punches, drills, awls and needles—Milwaukee Public Museum collection.

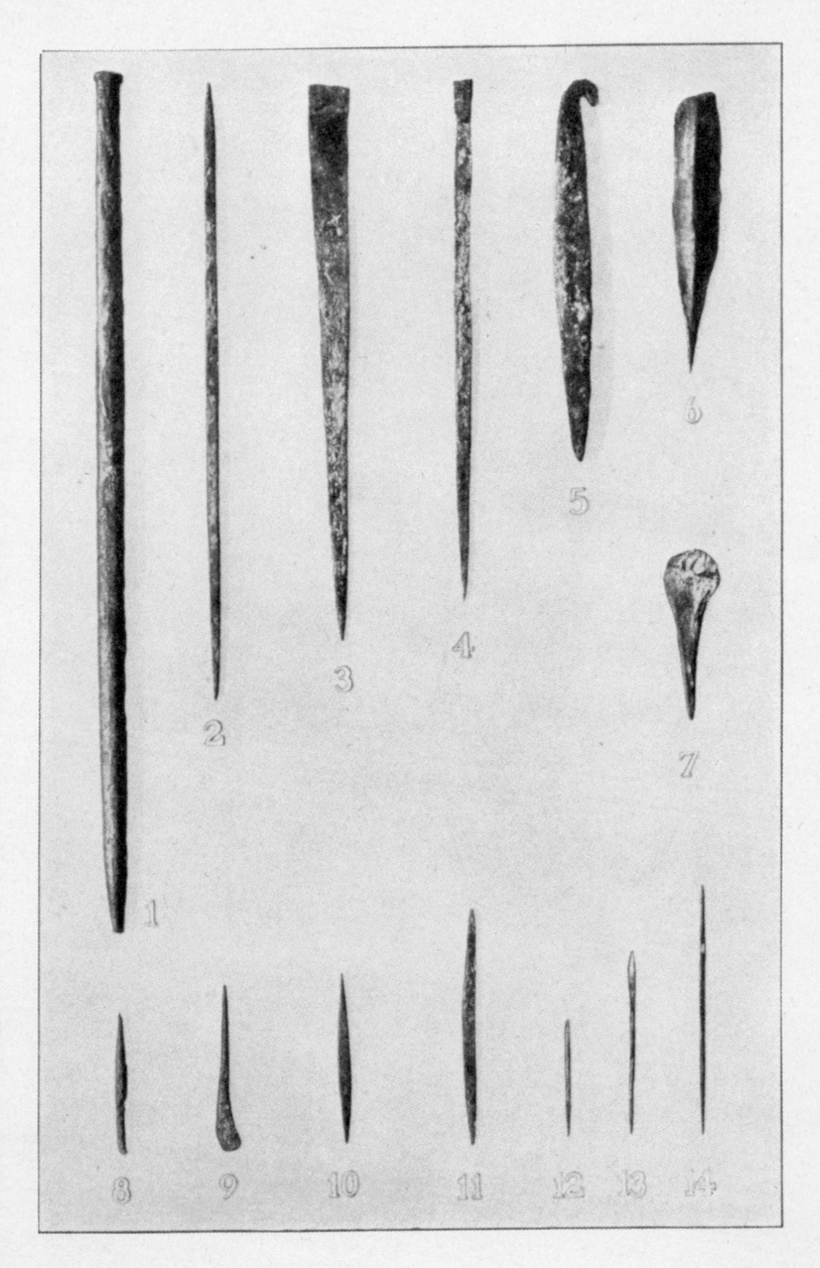
1
2
3
4
5
6
7
8
9
10
11
12
13
14

EXPLANATION OF PLATE XXVI.

Types of copper harpoons and fish-hooks—Milwaukee Public Museum collection.

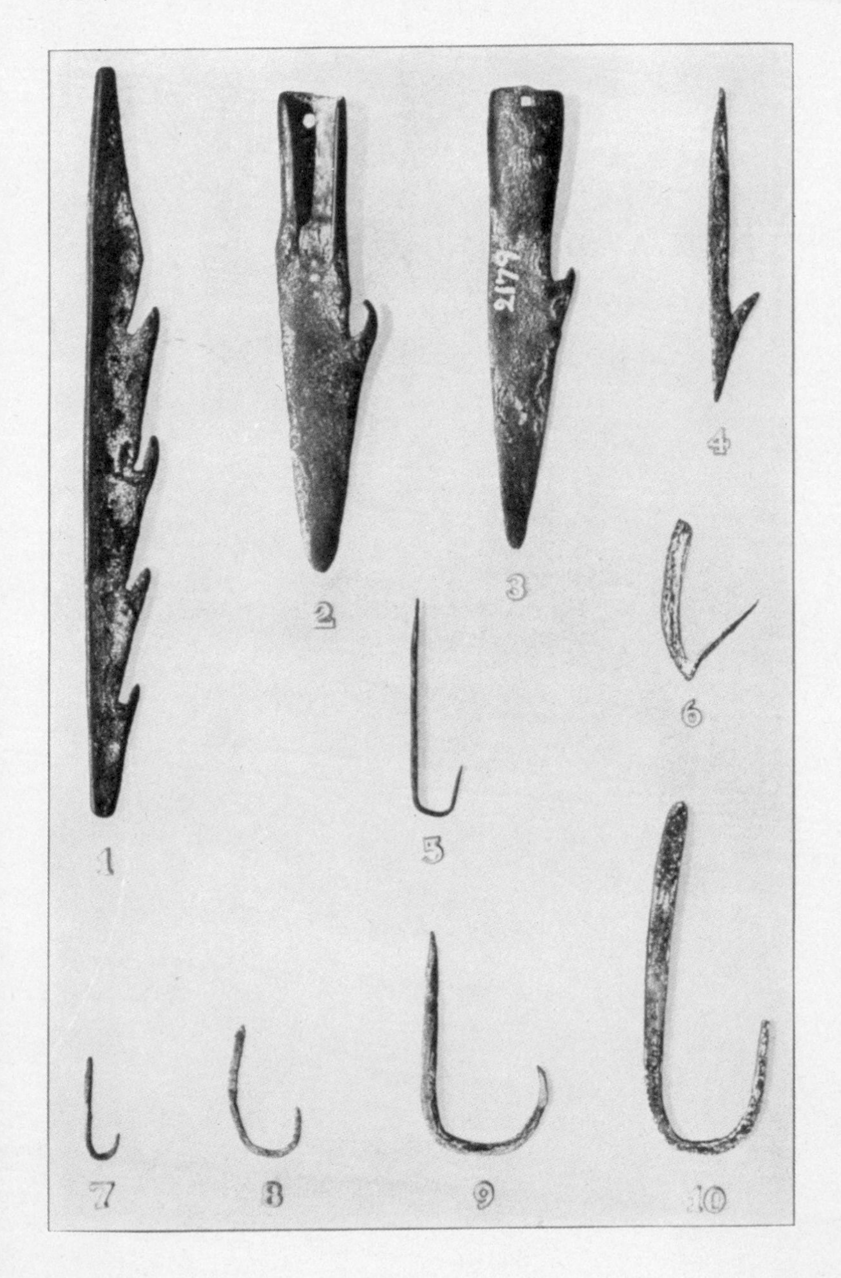
2179
1
2
3
4
5
6
7
8
9
10

EXPLANATION OF PLATE XXVII.

Types of copper crescents and spatulas—Milwaukee Public Museum collection.

1
2
3
4
5
6
7
8

EXPLANATION OF PLATE XXVIII.

Unique copper objects—Milwaukee Public Museum collection.

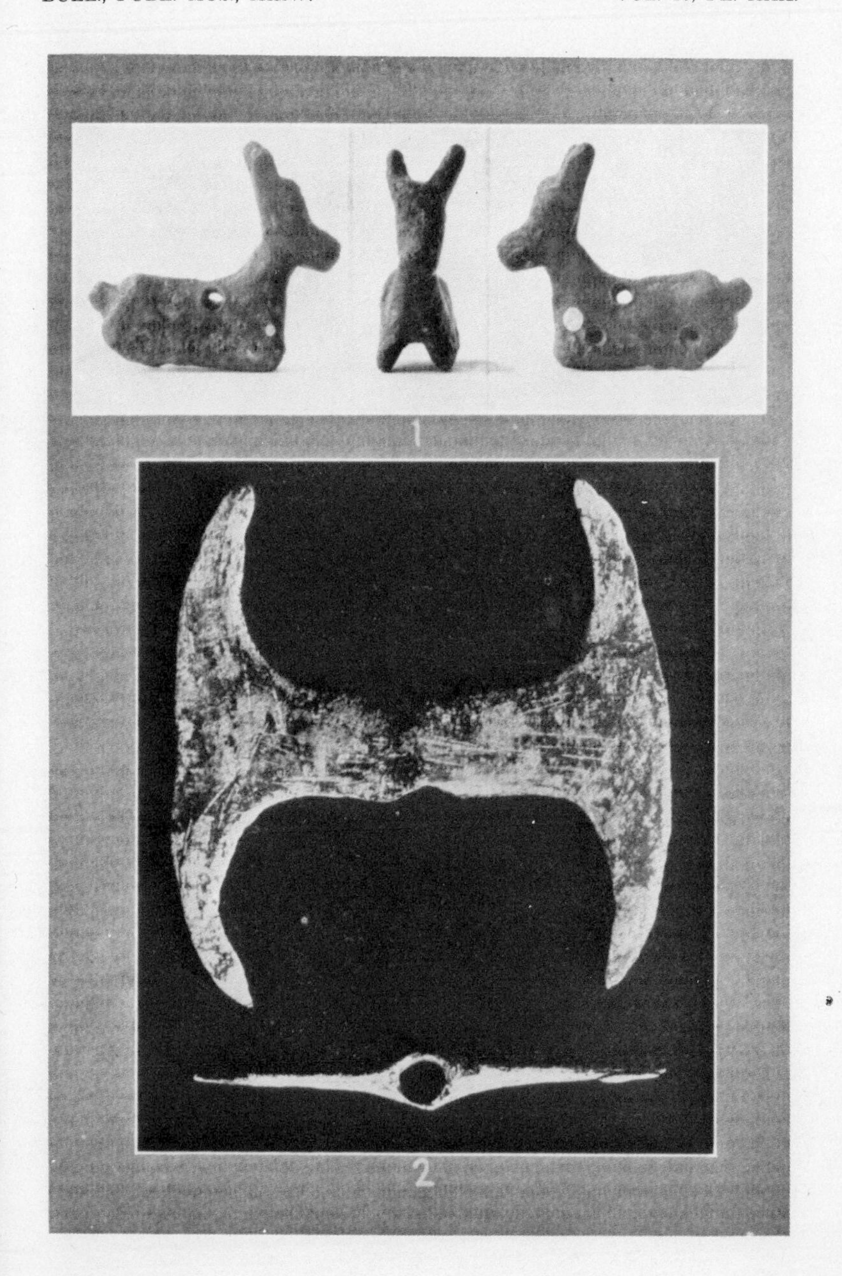
1
2